BASIC INCOME, UNEMPLOYMENT AND COMPENSATORY JUSTICE

T0189483

Basic Income, Unemployment and Compensatory Justice

by

Loek Groot

University of Amsterdam,
The Netherlands

WITH AN INTRODUCTORY CHAPTER BY PHILIPPE VAN PARIJS

KLUWER ACADEMIC PUBLISHERS
BOSTON / DORDRECHT / LONDON

A C.I.P. Catalogue record for this book is available from the Library of Congress.

ISBN   978-1-4419-5251-6 (PB)
ISBN   978-1-4020-2726-0 (e-book)

Published by Kluwer Academic Publishers,
P.O. Box 17, 3300 AA Dordrecht, The Netherlands.

Sold and distributed in North, Central and South America
by Kluwer Academic Publishers,
101 Philip Drive, Norwell, MA 02061, U.S.A.

In all other countries, sold and distributed
by Kluwer Academic Publishers,
P.O. Box 322, 3300 AH Dordrecht, The Netherlands.

Printed on acid-free paper

To Simon

TABLE OF CONTENTS

DETAILED TABLE OF CONTENTS

ACKNOWLEDGEMENTS

This book owes more debts than I can remember, let alone acknowledge. The discussion of basic income in this book has benefited greatly from two outstanding scholars: Philippe Van Parijs and Robert-Jan Van der Veen. There are good reasons for the choice of these two in particular. Philippe Van Parijs is not only my most cited author and the leading advocate of basic income since many years, but also the driving force behind the Basic Income European Network (B.I.E.N.) and the Charles Fourier Archive, which contains almost all publications about basic income. I am happy that his piece *Basic Income for All* is included in this volume as Introductory chapter. Robert-Jan Van der Veen is an expert about basic income for more than twenty years. In the course of writing this book, his friendly counsel, constructive criticism, challenging queries and encouragement were indispensable. Thanks to his comments, it became much clearer what I was trying to say. I wish to express profound gratitude to Yoe Brenner for his able assistance, unqualified support and open-mindedness during the first phase of writing this book. Marga Peeters and Lex Borghans kindly, enthusiastically and effectively helped me with many problems that bothered me for too long. In the final stage, Evelyn de Wijs skillfully helped me preparing the manuscript. Finally, my wife Tanja, although she did not type nor read any part, has contributed in innumerable ways and has been my support throughout.

INTRODUCTION

The Basic Income European Network (B.I.E.N.) defines a basic income (hereafter abbreviated as BI) as an income unconditionally granted to all on an individual basis, without means test or work requirement. In its pure form, a BI is equivalent to a negative income tax (see chapter 4). The main difference, apart from that Europeans are more inclined to discuss BI and Americans negative income tax, is that the former is paid *ex ante*, whereas the latter is paid *ex post*. In this book I want to put forward some arguments why the proposal of a BI (or negative income tax) is worth taking seriously as an alternative to the present, conditional, schemes of social security in force in all modern capitalist welfare states. The main claim is that the more serious the problem of (longlasting and large scale) involuntary unemployment, the more attractive and relevant the idea of BI. If the economy would ever be in a state of (near) full-employment, the main arguments given in this book in favour of an unconditional and substantial BI would not pertain. If the labour market would always clear (no involuntary unemployment) and everyone would have equal earning capacities, then everyone has the same access to jobs and the same opportunity to convert leisure time into money income by means of paid work. Consequently, the case for a BI would be weak, if not absent. Now suppose talent-based earning capacities are unequal, but keeping the assumption of full employment. In that case, taxation and redistribution measures to correct unequal earning capacities are warranted, but this can be done without implementing a BI, e.g. through progressive taxation and providing all kinds of subsidies and compensations to the least advantaged.[1] However, if there is structural unemployment (jobs shortage), I believe the most satisfactory way to deal with it is to implement a BI at a substantial level. However, doing this would be opposite to the prevailing policy response to unemployment, which is by and large to tighten the work- and means-tests for receiving social benefits and a gradual retrenchment of the welfare state, motivated by the hope to revert to a situation of full employment. In Figure 1, the present tendency is a movement from the top corner on the left to the right, while I advocate a vertical movement towards a BI.

Figure 1 concentrates on the treatment of social assistance claimants, those without social insurances who can only qualify for non-insurance based transfer payments if they can pass the eligibility conditions (for a more extensive comment on Figure 1, see Groot and Van der Veen, 2000a: 17-22). Although there has been an enormous expansion in the coverage of contributory insurance, still a significant part of the labour force has to rely on social assistance. This group comprises those with the weakest attachment to the labour market (e.g. single parent families with children, graduates just leaving school, and all welfare recipients who have

---

[1] Even in the situation of full employment, a BI might still be justified. Van Parijs (1995, chapter 4) argues for a BI financed by taxation of unequal wealth and job rents. Vandenbroucke (2001), using a framework in which the labour market invariably clears, shows that a BI can be defended if the government upholds a conception of the good life in which non-paid work and leisure is valued highly relative to paid work.

exceeded the maximum entitlement duration of social insurance benefits) and therefore are the most likely to be confronted with poverty and social exclusion.

Figure 1. Four ideal types of basic social security.

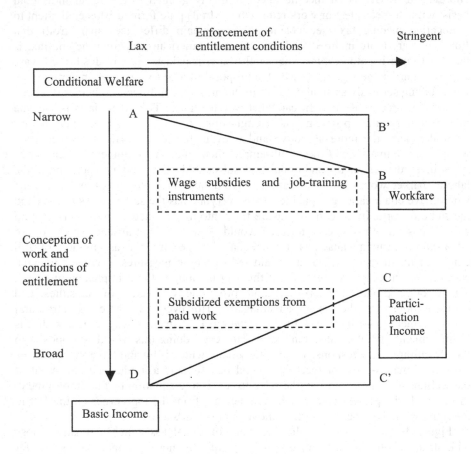

Moving from the left to the right along the horizontal dimension, the work-test conditions attached to receiving social assistance are tightened and thereby eliminate more and more the possibility of passive enjoyment of benefits. The vertical dimension represents the conception of work applied in the work test: this can be either narrow, with an exclusive emphasis on paid work, preferably in the private sector, or broad, where e.g. also volunteer and care work passes. On the corners of the rectangle, the idealtypes of social security are represented. The interrupted rectangles pictures the most obvious policy instruments that can be used by the government to reduce unemployment.

Under conditional welfare, the main instrument used by the government to reduce welfare dependency and social exclusion and to revert to full employment is to provide wage subsidies in order to enlarge the number of low paid jobs available for welfare recipients.[2] The more successful the wage subsidy policy is in expanding employment opportunities, the more reasonable it seems to impose the duty on welfare recipients to accept (common, accepted work instead of suitable) work and to apply sanctions if they do not cooperate. In moving from conditional welfare to workfare, the government takes a more active attitude, ranging from organized job-search and counseling activities, job placement services, public sector-sponsored training programmes to public and private job programmes. The design of work requirements in the configuration of workfare policies is motivated by promoting the work ethic, by improving skills and to discourage would-be welfare applications. This shift towards a more comprehensive active labour market policy allows a further tailoring of benefit rights and participation duties. The more active role of government coincides with steadily increasing demands on welfare recipients to make them 'deserving claimants', which also implies insertion into all kinds of governement-subsidized jobs (jobs which otherwise would not have existed, which explains why B' is below B). At the far right, the social benefit can be seen as an income derived from obligatory work (in conformity with the slogans 'only work should pay', 'work first' and 'work, not welfare'). The 'welfare for work' criterium is then applied narrrowly, where one is eligible if and only if one is at least engaged in a training or job programme preparing for regural paid work. To make workfare really work,[3] the government must therefore act as employer of last resort, which calls for a much larger task than anything European governments (except Sweden in the past) has started yet.[4]

As can be seen from Figure 1, the move towards activation and the strengthening of the conditionalities of social benefits, aimed at the build-up of marketable skills and, in the end, participation in paid work, is not the only possible policy response to unemployment. In principle, one can imagine a workfare scheme that is not biased,

---

[2] Although there are numerous public employment programmes and uncoordinated constituency-based training programmes in most European countries, they are predominantly temporary expedients and targeted to the far end of the labour queue. In other words, they fall short of a permanent and comprehensive workfare scheme.

[3] Solow (1998, 40-41) notes that the replacement of the welfare by workfare requires 'purposeful creation of jobs, in numbers, places, and forms that are suitable for the people who will fill them, and that can provide the sort of experience that may eventually have cash value in the open labor market. Any scheme that will do the trick will be costly, in budgetary dollars and in the need to invent and to staff institutions of a kind for which we have little experience or even intuition. The task is even harder than it sounds, because it involves swimming against the current. There has been in recent years a huge shift in demand away from unskilled labor. The source appears to have been mostly technological but the source is less important than the fact, and the fact suggests that the labor market will not naturally welcome an influx of unskilled workers.'

[4] Weir et al. (1988, 438-9) point out there is an inherent conflict in the workfare strategy: '... greatly increased monitoring capacity would still be required to oversee substantial new expenditures and enforce the (frequently unpleasant) work requirements. Increasing the authority of the government to regulate the conduct of citizens inevitably means building up the size and power of the state. That conservative supporters of workfare whose main concern is to deter participants would willingly go all the way with such a massive state builup is doubtful; that they could do so without inspiring political backlashes is even more dubious.'

on what counts as work, to paid work. In such a workfare scheme, participation in some kind of activity is mandatory, but the list of approved activities contains job and training opportunities preparing for regular paid work as well as activities belonging to the domain of unpaid work (e.g. care work). The broader and all-embracing the list of unpaid work activities, the more workfare approaches a participation income as defined by Atkinson (1993; 1996). However, even if the list of unpaid work activities is identical, there is still the difference that workfare is means-tested whereas participation income is non-means tested (see below). For instance, exempting single parent families with young children from the work-test while on welfare can be seen as giving them a participation income, but maintaining the means-test (and the accompanying poverty trap).

Like workfare, a participation income maintains the condition that a social contribution must be made in return for the benefit, but subscribing a relaxed view on work, it allows welfare recipients to choose among a wide range of approved activities (all kinds of care work, schooling, volunteer work, sabbatical leaves, etc.) to pass the work test. Finally, the BI scheme deploys a broad stance on work because the government does not try to push people into either paid or unpaid work. Contrary to the three other schemes, what one tries to exact from citizens is here replaced by voluntariness.

Another way to illustrate that a BI is radically different is to look at the pairs residual versus institutional, selective versus universal and temporary versus permanent with respect to minimim income protection arrangements. The main characteristic of the welfare state is the effective guarantee to all citizens that basic needs will be satisfied and a minimal share in welfare will be provided. The rights and duties of citizens are legally defined and can be enforced as social policy is laid down in social legislation. The social legislation details legal regulations and provisions, and leads to a mixed or social market economy: social goals not automatically realized by a free market capitalist economy (i.e. goals concerning full employment, conditions of employment, education, and income protection) are accounted for by an intervening government through social legislation. Both a BI and the present welfare state is compatible with this general description of social policy, but BI, being permanent and unconditional does so in an onorthodox way. The difference reflects the divergent intents behind both systems. In the residual vision of social policy one assumes that the basic institutions of society (family, market) are able to provide the means for the satisfaction of basic needs. Social policy is only meant to intervene if these institutions do not come up to the mark. In the institutional vision of social policy the basic institutions of society are not able any longer to provide basic social security for all, and hence the government has to intervene. Implicit in the residual vision is the principle of self-reliance (see also chapter 1, section 2). In addition, social security provisions can be either selective or universal in nature. The principle of selectivity states that social security provisions are only meant for those who need them (e.g. the poor), or only for those who have paid for them (e.g. those with unemployment or disability insurances). To be eligible, one must usually pass a work- and means-test. Universal social security benefits are provided to all, irrespective of the income or wealth position of the claimant. Examples include state pensions and child benefits. Finally, most social security benefits are not only conditional, but also temporary. They are meant to be a temporary provision during periods of unemployment or illness. A BI scheme more reflects the institutional vision. The institution of the BI can be seen as providing a minimum floor below which nobody can sink. As well as being a permanent measure, a BI is universal rather than selective in nature.

All schemes of social security will generate their own effects. Indeed, the most important of these is how the (economic) behaviour of residents will be affected. For instance, a system which is too generous, either because benefits are too high or because the conditions of entitlement are too loose, runs the risk of not being sustainable in the long run. So a prerequisite for any new design of social security is its long-run sustainability. The sustainability of a BI regime, irrespective of its desirability, depends to a large extent on the economic behaviour it gives way to. As will be argued later on, the economic feasibility of a BI is a highly contentious issue and no definitive answers can be given.

Why even consider implementing a BI scheme when its feasibility cannot be taken for granted? Any answer to this question must encompass the policy objectives which can be met by engaging in such a scheme. In the BI literature,[5] one finds a long list of these objectives: an adequate (unconditional) minimum income guarantee or basic social security for all, spreading of paid and unpaid labour within and between sexes, spreading of paid work through the encouragement of part-time work, reduction of involuntary unemployment at the bottom end of the labour market, improved adjustment of social security arrangements to diverse labour market behaviour (jobs provided by employment agencies, jobs on call), elimination of the poverty trap, elimination of stigma attached to social security benefits, discouragement of moonlighting, simplification of the social security system and reduction of administrative costs in providing social security, facilitation of techno-logical improvement and innovation, improvement of conditions of employment, more equitable or fair distribution of income through its redistributional effects and an increase in the number of self-employed persons. Some even speak of a new paradigm of social security, something beyond the present welfare state.[6]

Aside from the dividend paid to every citizen in Alaska from the revenues of the Alaska Permanent Income Fund (see Brown and Thomas, 1994) BI is still largely an imaginary construct. No welfare state has ever implemented a BI. It would mean a major break with the Beveridge-like social security systems now in force in most welfare states. Moreover, the study of the BI proposal is still surrounded by radical uncertainties with respect to the changes in citizens' patterns of behaviour in response to this change in social security. No firm conclusions can be drawn from economic models that try to scrutinize the effects of a substantial BI. Findings derived from local field experiments with a BI or a negative income tax are also of

---

[5] See e.g. de Beer (1987), Van Parijs (1991; 1995), Van der Veen (1991), Purdy (1990), Nooteboom (1993), Fitzpatrick (1999) and Van der Veen and Groot (2000).
[6] For instance, Van Parijs (1995, 64ff.) uses the term Painean justice. Roebroek and Hogenboom (1990) see the BI scheme as belonging to a new 'social-ecological paradigm' of social security and Goodin (2001) uses the term 'the post-productive welfare state'.

limited value to assess the viability of a BI scheme because the economy in which the experiment is situated has not been changed accordingly. However, even if economic models and experiments would show either the feasibility or the infeasibility of a substantial BI, and even if the BI proposal would never come high on the political agenda, the study of this alternative institutional framework may enable us to gain some insight in the direction in which present arrangements have to be adjusted. There is an immense grey area between a work- and learnfare scheme on the one extreme and a pure full BI scheme on the other. As long as the welfare state remains in difficulties, the BI scheme, even if not implemented, is still of interest. A more adequate understanding of what can be expected from a BI scheme in terms of social and economic outcomes may help us to evaluate the pros and cons and (dis)advantages of present welfare arrangements.

Creedy (1996, 58) says the BI scheme '... combines the simplest possible income tax with the simplest possible transfer system'. Van der Veen (1991) and Van Parijs (1995), each in their own way, argue that a BI scheme, pitched at the highest feasible level, is warranted within liberal-egalitarian justice, that is, it is under appropriate conditions, the most just scheme. Taken together, the BI scheme is both the *most simple* and the *most just* scheme, and perhaps also economically sustainable. Is this too good to be true? In the Introductory chapter, Philippe van Parijs gives a brief review of the main arguments and deals with basic income's historical pedigree, its close alternatives - negative income tax and stakeholder grants - and discusses competing policy instruments to fight unemployment - the Earned Income Tax Credit (EITC), wage subsidies and working time reduction.

The first chapter confronts the idea of BI with popular ideas of justice to which it does not fit in nicely on first glance. In this respect, it can be considered as related to the question of whether such a simple scheme is (politically and, to a lesser extent, economically) feasible. It starts from the fact that a majority of the public and also of professional economists support workfare-oriented schemes much more than its direct counterpart, an unconditional, no strings attached minimum income guarantee in the form of a BI or negative income tax. The predominant reason for the strong opposition to the BI proposal (and the majority support for workfare) is that it disconnects (strengthens) the link between paid work and income. Regarding the link between paid work and income, the conflict between the BI proposal and three popular ideas of justice is explored. First, one may find the proposal to provide an unconditional income guarantee as a matter of right to all citizens unattractive because it might harm the degree of self-reliance or self-support. Secondly, the opposition to BI can be motivated by arguing that its unconditional nature violates a widespread notion of reciprocity: one has to do something in return for receiving social benefits, whereas the BI is prone to the reciprocity-based parasitism objection. Finally, at first glance the BI does not fit with either the perfectionist or neutral work ethic. I argue that the conflict between BI and the notion of self-reliance is not as serious as it may seem at first sight. Although the BI removes the necessity to be self-reliant, it also offers better incentives and chances for those who are now on welfare to become self-reliant. The perfectionist work ethic is rejected because a liberal government should not impose a particular conception of the good life on its citizens. The parasitism objection (which may also be subsumed under the neutral work ethic) appears to be the strongest argument against the idea of BI.

The next two chapters address the claim of justice and attempts to investigate whether there is a justification for an unconditional BI if there is considerable unemployment. The specific question addressed in chapter 2 is whether the cause of compensatory justice is better served by a BI compared to conditional transfer payments. Before this question could be answered, three different approaches to compensatory justice are discussed. The objective and balancing approach are criticized either because they are impractical or run counter prohibitively high efficiency losses. The economist's approach to compensatory justice is the most fruitful, but it is incomplete in as far this approach does not pay sufficient attention to the conditions required to achieve compensatory justice. The conditions as given by a list are the competitive operation of the labour market, free career choices and full employment. Unemployment is the primary threat to compensatory justice. It is in the circumstance of unemployment that a substantial BI, providing all a socially acceptable alternative and a readily accessible fall-back without distorting the competitive forces within the labour market, is superior to a conditional scheme of minimum income support. Moreover, absence of parasitism (e.g., misuse of social benefits) and compensatory justice are mutually conflicting ideals in times of unemployment. To reduce improperly use of social benefits (and therefore parasitism) requires more severe means- and work-tests to become entitled to these benefits, but this also reduces bargaining powers of low wage workers towards potential employers. The unconditional BI, if it is set at a substantial level, precludes coercive pressure on (unemployed) workers to take jobs at rates that do not reflect compensatory justice. A high BI, ideally at a level close to the prevalent social minimum, makes the choice of unemployment acceptable and exerts an upward pressure on compensatory wage differentials necessary to induce workers to do unattractive work. Taking also into account that a BI scheme is redistributive from high wage to low wage workers, the main rationale for a BI in this respect is to improve the position of low wage workers.

Chapter 3 discusses the link between BI and un(der)employment, departing from a thought-experiment of an economy struck by unemployment with a market where scarce job rights can freely be traded. It turns out that the unemployment benefit derived from selling one's job rights to willing workers under the Labour Rights' framework and the BI are equivalent. So the advantages of the proposal to let the level of social benefits in times of unemployment to be determined by the market in job rights equally apply to the BI scheme. The main conclusions of this chapter are that there is no justification for a BI if there is no job scarcity (unemployment) at all. However, with unemployment, the level of the justified BI is higher, the higher the level of unemployment. Also, the level of the tax rate required to finance this BI varies proportionally with the level of unemployment because a higher tax rate is required to reduce the total labour supply to the number of jobs available. In this light, the tax rate can be interpreted as the price that full-time workers have to pay to appropriate scarce job rights. On the same footing, the level of BI can be interpreted as the (market) compensation for giving up one's equal share of jobs rights. The analysis further shows that under the criterium of maximin equality the level of justified BI may fall short of the maximum sustainable BI.

Taking the outcomes of both chapters together amounts to the following. There is a justification for a BI under unemployment, at a level which is higher, the higher the shortage of jobs (chapter 3). If the shortage of jobs is only modest, the argumentation in favour of a *substantial* BI, at least equal to the prevailing poverty level or social minimum, hinges on the potential of such a BI to facilitate compensatory justice on the labour market. A substantial BI, close to the level of the present social minimum, is the best guarantee that the demands of compensatory justice will be met in the circumstance of unemployment (chapter 2).

One will search in vain for a definite answer on the question whether a substantial BI is feasible or not. This question is not answerable, given the present state of economic science. Based on current theory and empirical research something can be said about which kinds of effects will result, specified by groups (e.g. low wage earners, high wage earners, welfare recipients), but not or hardly anything about the total effects of such a major change of conditional to unconditional social security. Bearing in mind that there is even no consensus among economists on the economic effects of a small change in the level the statutory minimum wage, it can safely be taken for granted that economists are not able to make a reliable prediction of the total impact of the many effects which would be brought about by the implementation of a BI. Taking this radical uncertainty surrounding the feasibility of the BI proposal into account, I will argue to repeat (prefereably in Europe) the negative income experiments held in the USA in the late 1960s and early 1970s. The limitations of a real life experiment are entirely different from the limitations of economic models. The insights obtained from the study of changes in behaviour of experimentals may thus constitute a complementary source of information compared to that obtained from economic models on the viability of BI. In chapter 4 it is argued that there are indeed good reasons to put the BI to the test by means of a real life experiment for a limited period, rather than calculating the economic feasibility of BI by a costing analysis in an economic model. Probably the most important one is that BI as a blueprint for social security reform is quite radical. Therefore it has to prove itself first of all as a viable alternative. The best way to do this is by means of a real life experiment. Such an experiment might also show which variant is the most promising. The criteria used for the choice of groups to be included in the experiment are twofold. First, the emphasis of the experiment is to research the labour supply effects of the groups for which there is the greatest disagreement among labour economists about expected negative labour supply responses and which are of great importance for the feasibility of a BI. These groups are the social assistance recipients and the low wage workers. Second, the choice is influenced by the desirability to minimize the cost of the experiment. Given the total budget for the experiment, the lower its cost per participant, the higher the number of participants and the longer can be the duration of the experiment. For this reason, a group of workers is included in the experiment who would not experience a change in net income if they take part in the experiment. If the experimentals of these groups do not change their labour supply, no extra costs for the experiment are incurred. Extra costs only occur when they decide to work less.

Finally, before a (partial) BI is implemented some intermediary measures (i.e. reducing the poverty trap and lowering the gross minimum wage whilst maintaining the net minimum wage) can be carried out bringing the present gross-net earnings trajectory closer to that of a BI scheme. This is discussed in chapter 5. These measures can be taken independently whether one aims towards a BI or not and may provide useful information on whether the final transition to BI is feasible.

Although there are many more topics related to BI that deserve attention, I hope that these chapters provide the readers some reasons why a BI might be an attractive solution in case that welfare states remain in trouble and what could be done to go from here to there.

A BASIC INCOME FOR ALL: A BRIEF DEFENCE *
TO SECURE REAL FREEDOM, GRANT EVERYONE
A SUBSISTENCE INCOME

Philippe Van Parijs

Entering the new millennium, I submit for discussion a proposal for the improvement of the human condition: namely, that everyone should be paid a universal basic income (UBI), at a level sufficient for subsistence.

In a world in which a child under five dies of malnutrition every two seconds, and close to a third of the planet's population lives in a state of "extreme poverty" that proves fatal to their health, the global enactment of such a basic income proposal may seem wildly utopian. Readers may suspect it to be impossible even in the wealthiest of OECD nations.

Yet, in those nations, productivity, wealth, and national incomes have advanced sufficiently far to support an adequate UBI. And if enacted, a basic income would serve as a powerful instrument of social justice: it would promote real freedom for all by providing the material resources that people need to pursue their aims. At the same time, it would help to solve the policy dilemmas of poverty and unemployment, and serve ideals associated with both the feminist and green movements. So I will argue.

I am convinced, along with many others in Europe, that – far from being utopian – a UBI makes common sense in the current context of the European Union.[1] And, as Brazilian senator Eduardo Suplicy has argued, it is also relevant to less-developed countries – not only because it helps keep alive the remote promise of a high level of social solidarity without the perversity of high unemployment, but also because it

* This introductory chapter by Philippe Van Parijs, co-founder and secretary of the Basic Income European Network, was written with primarily a North-American audience in mind and first published as the lead piece in the special issue of the *Boston Review* (25/5, October-November 2000) devoted to basic income. It was reprinted in *What's Wrong with a Free Lunch?* (J. Cohen & J. Rogers, eds., New Democracy Forum, Boston: Beacon Press, 2001), with a preface by Robert Solow and comments by Anne Alstott, Brian Barry, Claus Offe, Edmund S. Phelps, Herbert A. Simon, and others). Warm thanks to Josh Cohen, both for his contribution to the original version and for his permission to reprint a slightly adjusted version of the text in the present volume.
[1] Many academics and activists who share this view have joined the Basic Income European Network (BIEN). Founded in 1986, BIEN holds its tenth congress in Barcelona in September 2004, publishes an electronic newsletter (bien@basicincome.org), and maintains a web site that carries a comprehensive annotated bibliography in all EU languages (http://www.basicincome.org). For some useful European discussions, see Robert Jan Van der Veen and Loek Groot, eds., *Basic Income on the Agenda: Policy Objectives and Political Chances* (Amsterdam: Amsterdam University Press, 2000) and Angelika Krebs, ed., 'Basic Income?', special issue of *Analyse & Kritik* 22 (2), 2000, 153-269.

can inspire and guide more modest immediate reforms.[2] And if a UBI makes sense in Europe and in less developed countries, why should it not make equally good (or perhaps better) sense in North America?[3] After all, the United States is the only country in the world in which a UBI is already in place: in 1999, the Alaska Permanent Fund paid each person of whatever age who had been living in Alaska for at least one year an annual UBI of $1,680. This payment admittedly falls far short of subsistence, but has nonetheless become far from negligible two decades after its inception. Moreover, there was a public debate about UBI in the United States long before it started in Europe. In 1967, Nobel economist James Tobin published the first technical article on the subject, and a few years later, he convinced George McGovern to promote a UBI, then called 'demogrant', in his 1972 presidential campaign.[4]

To be sure, after this short public life the UBI has sunk into near-oblivion in North America. For good reasons? I believe not. There are many relevant differences between the United States and the European Union in terms of labor markets, educational systems, and ethnic make-up. But none of them makes the UBI intrinsically less appropriate for the United States than for the European Union. More important are the significant differences in the balance of political forces. In the United States, far more than in Europe, the political viability of a proposal is deeply affected by how much it caters to the tastes of wealthy campaign donors. This is bound to be a serious additional handicap for any proposal that aims to expand options for, and empower, the least wealthy. But let's not turn necessity into virtue, and sacrifice justice in the name of increased political feasibility. When fighting to reduce the impact of economic inequalities on the political agenda, it is essential, in the United States as elsewhere, to propose, explore, and advocate ideas that are ethically compelling and make economic sense, even when their political feasibility remains uncertain. Sobered, cautioned, and strengthened by Europe's debate of the last two decades, here is my modest contribution to this task.

UBI DEFINED

By universal basic income I mean an income paid by a government, at a uniform level and at regular intervals, to each adult member of society. The grant is paid, and its level is fixed, irrespective of whether the person is rich or poor, lives alone or with others, is willing to work or not. In most versions – certainly in mine – it is granted not only to citizens but to all permanent residents.

---

[2] Federal senator for the State of Sao Paulo and member of the Workers Party (PT), Eduardo Suplicy managed to get a basic income proposal approved by Brazil's federal Congress in 2003 and signed into law by President Lula da Silva in January 2004. It is meant to be implemented gradually as from 2005, 'starting with the neediest categories'.

[3] Two UBI networks were set up in the Spring of 2000: the United States Basic Income Guarantee Network, c/o Dr. Karl Widerquist (www.usbig.net); and Basic Income/Canada, c/o Prof. Sally Lerner (www.fes.uwaterloo.ca/Research/FW).

[4] See James Tobin, Joseph A. Pechman, and Peter M. Mieszkowski, 'Is a Negative Income Tax Practical?', *Yale Law Journal* 77 (1967), 1-27. See also a recent conversation with Tobin in BIEN's newsletter ('James Tobin, the Demogrant and the Future of U.S. Social Policy', in *Basic Income* 29 (Spring 1998), available on BIEN's web site).

The UBI is called 'basic' because it is something on which a person can safely count, a material foundation on which a life can firmly rest. Any other income – whether in cash or in kind, from work or savings, from the market or the state – can lawfully be added to it. On the other hand, nothing in the definition of UBI, as it is here understood, connects it to some notion of 'basic needs'. A UBI, as defined, can fall short of or exceed what is regarded as necessary to a decent existence.

I favor the highest sustainable such income, and believe that all the richer countries can now afford to pay a basic income above subsistence. But advocates of a UBI do not need to press for a basic income at this right away. In fact, the easiest and safest way forward, though details may differ considerably from one country to another, is likely to consist of enacting a UBI first at a level below subsistence, and then increasing it over time.

The idea of UBI is at least 150 years old. Its two earliest known formulations were inspired by Charles Fourier, the prolific French utopian socialist. In 1848, while Karl Marx was finishing off the Communist Manifesto around the corner, the Brussels-based Fourierist author Joseph Charlier published *Solution of the Social Problem*, in which he argued for a 'territorial dividend' owed to each citizen by virtue of our equal ownership of the nation's territory. The following year, John Stuart Mill published a new edition of his *Principles of Political Economy*, which contains a sympathetic presentation of Fourierism ('the most skillfully combined, and with the greatest foresight of objections, of all the forms of Socialism') rephrased so as to yield an unambiguous UBI proposal: 'In the distribution, a certain minimum is first assigned for the subsistence of every member of the community, whether capable or not of labour. The remainder of the produce is shared in certain proportions, to be determined beforehand, among the three elements, Labour, Capital, and Talent.'[5]

Under various labels ('state bonus', 'national dividend', 'social dividend', 'citizen's wage', 'citizen's income', 'universal grant', 'basic income', etc.) the idea of a UBI was repeatedly taken up in intellectual circles throughout the twentieth century. It was seriously discussed by left-wing academics such as G.D.H. Cole and James Meade in England between the World Wars and, via Abba Lerner, it seems to have inspired Milton Friedman's proposal for a 'negative income tax'.[6] But only since the late-1970s has the idea gained real political currency in a number of European countries, starting with the Netherlands and Denmark. A number of political parties, usually green or 'left-liberal' (in the European sense), have now made it part of their official party program.

UBI AND EXISTING PROGRAMS

To appreciate the significance of this interest and support, it is important to understand how a UBI differs from existing benefit schemes. It obviously differs

---

[5] See Joseph Charlier, *Solution du problème social ou constitution humanitaire* (Bruxelles: Chez tous les libraires du Royaume, 1848); John Stuart Mill, *Principles of Political Economy*, 2nd ed. [1849] (New York: Augustus Kelley, 1987).
[6] See the exchange between Eduardo Suplicy and Milton Friedman in *Basic Income* 34 (June 2000).

from traditional social-insurance-based income maintenance institutions (such as Social Security), whose benefits are restricted to wage workers who have contributed enough out of their past earnings to become eligible. But it also differs from Western European or North American conditional minimum-income schemes (such as welfare).

Many, indeed most, West European countries introduced some form of guaranteed minimum-income scheme at some point after World War II.[7] But these schemes remain conditional: to receive an income grant a beneficiary must meet three more or less stringent variants of the following three requirements: if she is able to work, she must be willing to accept a suitable job, or to undergo suitable training, if offered; she must pass a means test, in the sense that she is only entitled to the benefit if there are grounds to believe that she has no access to a sufficient income from other sources; whether she is entitled to a benefit and, if so, how high it is depends on her household situation – for example, whether she lives on her own, with a person who has a job, with a jobless person, etc. By contrast, a UBI does not require satisfaction of any of these conditions.

Advocates of a UBI may, but generally do not, propose it as a full substitute for existing conditional transfers. Most supporters want to keep – possibly in simplified forms and necessarily at reduced levels – publicly organized social insurance and disability compensation schemes that would supplement the unconditional income while remaining subjected to the usual conditions. Indeed, if a government implemented an unconditional income that was too small to cover basic needs – which, as I previously noted, would almost certainly be the case at first – UBI advocates would not want to eliminate the existing conditional minimum-income schemes, but only to readjust their levels.

In the context of Europe's most developed welfare states, for example, one might imagine the immediate introduction of universal child benefits and a strictly individual, noncontributory basic pension as full substitutes for existing means-tested benefit schemes for the young and the elderly. Indeed, some of these countries already have such age-restricted UBIs for the young and the elderly. Contributory retirement insurance schemes, whether obligatory or optional, would top up the basic pension.

As for the working-age population, advocates of a universal minimum income could, in the short term, settle for a 'partial' (less-than-subsistence), but strictly individual UBI, initially pitched at, say, half the current guaranteed minimum income for a single person. In US terms, that would be about $250 per month, or $3,000 a year. For households whose net earnings are insufficient to reach the socially defined subsistence level, this unconditional and individual floor would be supplemented by means-tested benefits, differentiated according to household size and subjected, as they are now, to some work requirements.

---

[7] The latest countries to introduce a guaranteed minimum income at national level were France (in 1988) and Portugal (in 1997). Out of the European Union's fifteen member states, only Italy and Greece have no such scheme.

UBI AND SOME ALTERNATIVES

While the UBI is thus different from traditional income maintenance schemes, it also differs from a number of other innovative proposals that have attracted recent attention. Perhaps closest to a UBI are various negative income tax (NIT) proposals.[8]

NIT

Though the details vary, the basic idea of a NIT is to grant each citizen a basic income, but in the form of a refundable tax credit. From the personal tax liability of each household, one subtracts the sum of the basic incomes of its members. If the difference is positive, a positive tax needs to be paid. If it is negative, a benefit (or negative tax) is paid by the government to the household. In principle, one can achieve exactly the same distribution of post-tax-and-transfer income among households with a UBI or with a NIT. Indeed, the NIT might be cheaper to run, since it avoids the to-and-fro that results from paying a basic income to those with a substantial income and then taxing it back.

Still, a UBI has three major advantages over a NIT. First, any NIT scheme would have the desired effects on poverty only if it was supplemented by a system of advance payments sufficient to keep people from starving before their tax forms are examined at the end of the fiscal year. But from what we know of social welfare programs, ignorance or confusion is bound to prevent some people from getting access to such advance payments. The higher rate of take-up that is bound to be associated with a UBI scheme matters greatly to anyone who wants to fight poverty.

Second, although a NIT could in principle be individualized, it operates most naturally and is usually proposed at the household level. As a result, even if the inter-household distribution of income were exactly the same under a NIT and the corresponding UBI, the intra-household distribution will be far less unequal under UBI. In particular, under current circumstances, the income that directly accrues to women will be considerably higher under the UBI than the NIT, since the latter tends to ascribe to the household's higher earner at least part of the tax credit of the low- or non-earning partner.

Third, a UBI can be expected to deal far better than NIT with an important aspect of the 'unemployment trap' that is stressed by social workers but generally overlooked by economists. Whether it makes any sense for an unemployed person to look for or accept a job does not only depend on the difference between income at work and out of work. What deters people from getting out to work is often the reasonable fear of uncertainty. While they try a new job, or just after they lose one, the regular flow of benefits is often interrupted. The risk of administrative time lags – especially among people who may have a limited knowledge of their entitlements and the fear of going into debt, or for people who are likely to have no savings to

---

[8] In the United States, one recent proposal of this type has been made by Fred Block and Jeff Manza, 'Could We End Poverty in a Postindustrial Society? The Case for a Progressive Negative Income Tax', *Politics and Society* 25 (December 1997), 473-511.

fall back on – may make sticking to benefits the wisest option. Unlike a NIT, a UBI provides a firm basis of income that keeps flowing whether one is in or out of work. And it is therefore far better suited to handle this aspect of the poverty trap.

The Stakeholder Society

UBI also differs from the lump-sum grant, or 'stake', that Thomas Paine and Orestes Brownson – and, more recently, Bruce Ackerman and Anne Alstott – have suggested be universally awarded to citizens at their maturity in a refashioned 'stakeholder society'.[9] Ackerman and Alstott propose that, upon reaching age 21, every citizen, rich or poor, should be awarded a lump-sum stake of $80,000. This money can be used in any way its recipient wishes – from investing in the stock market or paying for college fees to blowing it all in a wild night of gambling. The stake is not conditioned on recipients being 'deserving', or having shown any interest in contributing to society. Funding would be provided by a 2 percent wealth tax, which could be gradually replaced over time (assuming a fair proportion of recipients ended their lives with enough assets) by a lump-sum estate tax of $80,000 (in effect requiring the recipient to pay back the stake).

I am not opposed to a wealth or estate tax, nor do I think it is a bad idea to give everyone a little stake to get going with their adult life. Moreover, giving a large stake at the beginning of adult life might be regarded as formally equivalent – with some freedom added – to giving an equivalent amount as a life-long unconditional income. After all, if the stake is assumed to be paid back at the end of a person's life, as it is in the Ackerman/Alstott proposal, the equivalent annual amount is simply the stake multiplied by the real rate of interest, say an amount in the (very modest) order of $2,000 annually, or hardly more than Alaska's dividend. If instead people are entitled to consume their stake through life – and who would stop them? – the equivalent annual income would be significantly higher.

Whatever the level, given the choice between an initial endowment and an equivalent life-long UBI, we should go for the latter rather than the former. Endowments are rife with opportunities for waste, especially among those less well equipped by birth and background to make good use of the opportunity the stake supplies. To achieve, on an ongoing basis, the goal of some baseline income maintenance, it would therefore be necessary to keep a means-tested welfare system, and we would be essentially back to our starting point – the need and desirability of a UBI as an alternative to current provisions.

---

[9] See Bruce Ackerman and Anne Alstott, *The Stakeholder Society* (New Haven: Yale University Press, 1999). See also, for a discussion, Wright, Erik O., ed., *Redesigning Distribution: Basic Income and Stakeholder Grants as Cornerstones of a More Egalitarian Capitalism*, special issue of *Politics & Society* 32 (1), March 2004, and Dowding, Keith, De Wispelaere, Jurgen & White, Stuart, eds., *The Ethics of Stakeholding*, Houndmills: Palgrave Macmillan, 2003. Their proposal is a sophisticated and updated version of a proposal made by Thomas Paine to the French Directoire, 'Agrarian justice' [1796], in *The Life and Major Writings of Thomas Paine*, P.F. Foner, ed., (Secaucus , N.J.: Citadel Press, 1974), 605-623. A similar program was proposed, independently, by the New England liberal, and later arch-conservative, Orestes Brownson in the *Boston Quarterly Review* of October 1840. If the American people are committed to the principle of 'equal chances', he argued, then they should make sure that each person receives, on maturity, an equal share of the 'general inheritance'.

WHY A UBI?

So much for definitions and distinctions. Let us now turn to the central case for UBI.

Justice

The main argument for UBI is founded on a view of justice. Social justice, I believe, requires that our institutions be designed to best secure *real freedom* to all.[10] Such a 'real-libertarian' conception of justice combines two ideas. First, the members of society should be formally free, with a well-enforced structure of property rights that includes the ownership of each by herself. What matters to a real libertarian, however, is not only the protection of individual rights, but assurances of the real value of those rights: we need to be concerned not only with liberty, but, in John Rawls's phrase, with the 'worth of liberty'. At first approximation, the worth or real value of a person's liberty depends on the resources the person has at her command to make use of her liberty. So it is therefore necessary that the distribution of opportunity – understood as access to the means that people need for doing what they might want to do – be designed to offer the greatest possible real opportunity to those with least opportunities, subject to everyone's formal freedom being respected.

This notion of a just, free society needs to be specified and clarified in many respects.[11] But in the eyes of anyone who finds it attractive, there cannot but be a strong presumption in favor of UBI. A cash grant to all, no questions asked, no strings attached, at the highest sustainable level, can hardly fail to advance that ideal. Or if it does not, the burden of argument lies squarely on the side of the challengers.

Jobs and Growth

A second way to make the case for UBI is more policy-oriented. A UBI might be seen as a way to solve the apparent dilemma between a European-style combination of limited poverty and high unemployment and an American-style combination of low unemployment and widespread poverty. The argument can be spelled out, very schematically, as follows.

For over two decades, most West European countries have been experiencing massive unemployment. Even at the peak of the jobs cycle, millions of Europeans are vainly seeking work. How can this problem be tackled? For a while, the received wisdom was to deal with massive unemployment by speeding up the rate of growth.

---

[10] For a more developed argument, see Philippe Van Parijs, *Real Freedom for All* (New York: Oxford University Press, 1995) and, for a critical discussion, Reeve, Andrew & Williams, Andrew, eds., 2003. *Real Libertarianism Assessed. Political Theory after Van Parijs* (Basingstoke: Palgrave Macmillan).

[11] One can think of alternative normative foundations. For example, under some empirical assumptions a UBI is also arguably part of the package that Rawls's difference principle would justify. See, for example, Walter Schaller, 'Rawls, the Difference Principle, and Economic Inequality', *Pacific Philosophical Quarterly* 79 (1998), 368-91; Philippe Van Parijs, 'Difference Principles', in *The Cambridge Companion to John Rawls*, Samuel Freeman, ed., (Cambridge: Cambridge University Press, 2002, 200-240). Alternatively, one might view a UBI as a partial embodiment of the Marxian principle of distribution according to needs. See Robert J. Van der Veen and Philippe Van Parijs, 'A Capitalist Road to Communism', *Theory and Society* 15 (1986), 635-55.

But considering the speed with which technological progress was eliminating jobs, it became apparent that a fantastic rate of growth would be necessary even to keep employment stable, let alone to reduce the number of unemployed. For environmental and other reasons, such a rate of growth would not be desirable. An alternative strategy was to consider a substantial reduction in workers' earnings. By reducing the relative cost of labor, technology could be redirected in such a way that fewer jobs were sacrificed. A more modest and therefore sustainable growth rate might then be able to stabilize and, gradually, reduce present levels of unemployment. But this could only be achieved at the cost of imposing an unacceptable standard of living on a large part of the population, all the more so because a reduction in wages would require a parallel reduction in unemployment benefit and other replacement incomes, so as to preserve work incentives.

If we reject both accelerated growth and reduced earnings, must we also give up on full employment? Yes, if by full employment we mean a situation in which virtually everyone who wants a full-time job can obtain one that is both affordable for the employer without any subsidy and affordable for the worker without any additional benefit. But perhaps not, if we are willing to redefine full employment by either shortening the working week, paying subsidies to employers, or paying subsidies to employees.

A first option, particularly fashionable in France at the moment, consists in a social redefinition of 'full time' – that is, a reduction in maximum working time, typically in the form of a reduction in the standard length of the working week. The underlying idea is to ration jobs: because there are not enough jobs for everyone who would like one, let us not allow a subset to appropriate them all.

On closer scrutiny, however, this strategy is less helpful than it might seem. If the aim is to reduce unemployment, the reduction in the work week must be dramatic enough to more than offset the rate of productivity growth. If this dramatic reduction is matched by a proportional fall in earnings, the lowest wages will then fall – unacceptably – below the social minimum. If, instead, total earnings are maintained at the same level, if only for the less well paid, labor costs will rise. The effect on unemployment will then be reduced, if not reversed, as the pressure to eliminate the less-skilled jobs through mechanization is stepped up. In other words, a dramatic reduction in working time looks bound to be detrimental to the least qualified jobs – either because it kills the supply (they pay less than replacement incomes) or because it kills the demand (they cost firms a lot more per hour than they used to).

It does not follow that the reduction of the standard working week can play no role in a strategy for reducing unemployment without increasing poverty. But to avoid the dilemma thus sketched, it needs to be coupled with explicit or implicit subsidies to low-paid jobs. For example, a reduction of the standard working week did play a role in the so-called 'Dutch miracle' – the fact that, in the last decade or so, jobs expanded much faster in the Netherlands than elsewhere in Europe. But this was mainly as a result of the workers' standard working week falling below the firms' usual operating time and thereby triggering off a restructuring of work organization that involved far more part-time jobs. But these jobs could not have developed without the large implicit subsidies they enjoy, in the Netherlands, by virtue of a universal basic pension, universal child benefits, and a universal health-care system.

Any strategy for reducing unemployment without increasing poverty depends, then, on some variety of the 'active' welfare state – that is, a welfare state that does not subsidize passivity (the unemployed, the retired, the disabled, etc.) but systematically and permanently (if modestly) subsidizes productive activities. Such subsidies can take many different forms. At one extreme, they can take the form of general subsidies to employers at a level that is gradually reduced as the hourly wage rate increases. Edmund Phelps has advocated a scheme of this sort, restricted to full-time workers, for the United States.[12] In Europe, this approach usually takes the form of proposals to abolish employers' social security contributions on the lower earnings while maintaining the workers' entitlements to the same level of benefits.

At the other extreme we find the UBI, which can also be understood as a subsidy, but one paid to the employee (or potential employee), thereby giving her the option of accepting a job with a lower hourly wage or with shorter hours than she otherwise could. In between, there are a large number of other schemes, such as the US Earned Income Tax Credit, or various benefit schemes restricted to people actually working or actively looking for full-time work.

A general employment subsidy and a UBI are very similar in terms of the underlying economic analysis and, in part, in what they aim to achieve. For example, both address head-on the dilemma mentioned in connection with reductions in work time: they make it possible for the least skilled to be employed at a lower cost to their employer, without thereby impoverishing workers.

The two approaches are, however, fundamentally different in one respect. With employer subsidies, the pressure to take up employment is kept intact, possibly even increased; with a UBI, that pressure is reduced. This is not because permanent idleness becomes an attractive option: even a large UBI cannot be expected to secure a comfortable standard of living on its own. Instead, a UBI makes it easier to take a break between two jobs, reduce working time, make room for more training, take up self-employment, or to join a cooperative. And with a UBI, workers will only take a job if they find it suitably attractive, while employer subsidies make unattractive, low-productivity jobs more economically viable. If the motive in combating unemployment is not some sort of work fetishism – an obsession with keeping everyone busy – but rather a concern to give every person the possibility of taking up gainful employment in which she can find recognition and accomplishment, then the UBI is to be preferred.

Feminist and Green Concerns

A third piece of the argument for UBI takes particular note of its contribution to realizing the promise of the feminist and green movements. The contribution to the first should be obvious. Given the sexist division of labor in the household and the

---

[12] See Edmund S. Phelps, *Rewarding Work* (Cambridge, Mass.: Harvard University Press, 1997).

special 'caring' functions that women disproportionately bear, their labor market participation, and range of choice in jobs, is far more constrained than those of men. Both in terms of direct impact on the inter-individual distribution of income and in terms of the longer-term impact on job options, a UBI is therefore bound to benefit women far more than men. Some of them, no doubt, will use the greater material freedom UBI provides to reduce their paid working time and thereby lighten the 'double shift' at certain periods of their lives. But who can sincerely believe that working subject to the dictates of a boss forty hours a week is a path to liberation? Moreover, it is not only against the tyranny of bosses that UBI supplies some protection, but also against the tyranny of husbands and bureaucrats. It provides a modest but secure basis on which the more vulnerable can stand, as marriages collapse or administrative discretion is misused.

To discuss the connection between UBI and the green movement, it is useful to view the latter as an alliance of two components. Very schematically, the environmental component's central concern is with the pollution generated by industrial society, and its central objective is the establishment of a society that can be sustained by its physical environment. The green-alternative component's central concern, on the other hand, is with the alienation generated by industrial society; its central objective is to establish a society in which people spend a great deal of their time on 'autonomous' activities, ruled by neither the market nor the state. For both components, there is something very attractive in the idea of a UBI.

The environmentalists' chief foe is productivism, the obsessive pursuit of economic growth. And one of most powerful justifications for fast growth, in particular among the working class and its organizations, is the fight against unemployment. What the idea of a UBI provides, as argued above, is a coherent strategy for tackling the latter without relying on faster growth. The availability of such a strategy undermines the broad productivist coalition and thereby improves the prospects for realizing environmentalist objectives in a world in which pollution (even in the widest sense) is not the only thing most people care about.

Green-alternatives should also be attracted to basic income proposals, for a UBI can be viewed as a general subsidy financed by the market and state spheres to the benefit of the autonomous sphere. Part of this impact consists in the UBI giving everyone some real freedom – as opposed to a sheer right – to withdraw from paid employment in order to perform autonomous activities, such as grass-root militancy or unpaid care work. But part of the impact also consists in giving the least well endowed greater power to turn down jobs that they do not find sufficiently fulfilling, and in thereby creating incentives to design and offer less alienated employment.

To illustrate, consider two jobs paying equally low wages. One is in a dangerous food-processing plant, saturated with the stench of animal entrails, run under speed-up conditions and poisonous labor relations. The other is in a sunny, brightly painted day-care co-op that specializes in music and arts training for children and where executive decisions were made by consensus. Which job would you rather have? What a UBI would do is give individuals the ability to answer – the power to say 'no' to meaningless low-wage employment and to say 'yes' to socially important employment that is undervalued by the market. Through the natural subsequent workings of labor market forces, the result would be a general improvement in the conditions of low-wage work, and a likely shift in its composition closer to the green and alternative ideal of a humane society.

SOME OBJECTIONS

Suppose everything I have said thus far is persuasive: that the UBI, if it could be instituted, would be a natural and attractive way of ensuring a fair distribution of real freedom, fighting unemployment without increasing poverty, and promoting the central goals of both the feminist and the green movements. What are the objections?

Perhaps the most common is that a UBI would cost too much. Such a statement is of course meaningless if the amount and the scale is left unspecified. At a level of $150 per month and per person, a UBI is obviously affordable in some places, since this is the monthly equivalent of what every Alaskan receives as an annual dividend. Could one afford a UBI closer to the poverty line? By simply multiplying the poverty threshold for a one-person household by the population of a country, one soon reaches scary amounts – often well in excess of the current level of total government expenditure.

But these calculations are misleading. A wide range of existing benefits can be abolished or reduced once a UBI is in place. And for most people of working age, the basic income and the increased taxes (most likely in the form of an abolition of exemptions and of low tax rates for the lowest income brackets) required to pay for it will largely offset each other. In a country such as the United States, which has developed a reasonably effective revenue collection system, what matters is not the gross cost but its distributive impact – which could easily work out the same for a UBI or a NIT.

Estimates of the net budgetary cost of various UBI and NIT schemes have been made both in Europe and the United States.[13] Obviously, the more comprehensive and generous existing means-tested minimum-income schemes are, the more limited the net cost of a UBI scheme at a given level. But the net cost is also heavily affected by two other factors. Does the scheme aim to achieve an effective rate of taxation (and hence of disincentive to work) at the lower end of the distribution of earnings no higher than the tax rates higher up? And does it give the same amount to each member of a couple as to a single person? If the answer is positive on both counts, a scheme that purports to lift every household out of poverty has a very high net cost, and would therefore generate major shifts in the income distribution, not only from richer to poorer households, but also from single people to couples.[14] This

---

[13] In the US case, for example, a fiscally equivalent negative income tax scheme proposed by Fred Block and Jeff Manza, which would raise all base incomes to at least 90 percent of the poverty line (and those of poor families well above that), would, in mid-1990s dollars, cost about $60 billion annually.

[14] To fund this net cost, the personal income tax is obviously not the only possible source. In some European proposals, at least part of the funding comes from ecological, energy or land taxes; from a tax on value; or from non-inflationary money creation; or possibly even from Tobin taxes on international financial transactions (although it is generally recognized that the funding of a basic income in rich countries would not exactly be a priority in the allocation of whatever revenues may be collected from this source). But none of these sources could realistically enable us to dispense with personal income

does not mean that it is 'unaffordable', but that a gradual approach is required if sudden sharp falls in the disposable incomes of some households are to be avoided. A basic income or negative income tax at the household level is one possible option. A strictly individual, but 'partial' basic income, with means-tested income supplements for single adult households, is another.

A second frequent objection is that UBI would have perverse labor supply effects. (In fact, some American income maintenance experiments in the 1970s showed such effects.) The first response must be: 'So what?' Boosting the labor supply is no aim in itself. No one can reasonably want an overworked, hyperactive society. Give people of all classes the opportunity to reduce their working time or even take a complete break from work in order to look better after their children or elderly relatives. You will not only save on prisons and hospitals. You will also improve the human capital of the next generation. A modest UBI is a simple and effective instrument in the service of keeping a socially and economically sound balance between the supply of paid labor and the rest of our lives.

It is of the greatest importance that our tax-and-transfer systems should not trap the least skilled, or those whose options are limited for some other reason, in a situation of idleness and dependency. But it is precisely awareness of this risk that has been the most powerful factor in arousing public interest for a UBI in those European countries in which a substantial means-tested guaranteed minimum income had been operating for some time. It would be absurd to deny that such schemes depress in undesirable ways workers' willingness to accept low-paid jobs and stick with them, and therefore also employers' interest in designing and offering such jobs. But reducing the level or security of income support, on the pattern of the United States 1996 welfare reform, is not the only possible response. Reducing the various dimensions of the unemployment trap by turning means-tested schemes into universal ones is another. Between these two routes, there cannot be much doubt about what is to be preferred by people committed to combining a sound economy and a fair society – as opposed to boosting labor supply to the maximum.

A third objection is moral rather than simply pragmatic. A UBI, it is often said, gives the undeserving poor something for nothing. According to one version of this objection, a UBI conflicts with the fundamental principle of reciprocity: the idea that people who receive benefits should respond in kind by making contributions. Precisely because it is unconditional, it assigns benefits even to those who make no social contribution – who spend their mornings bickering with their partner, surf off Malibu in the afternoon, and smoke pot all night.

One response consists in simply asking: How many would actually choose this life? How many, compared to the countless people who spend most of their days doing socially useful but unpaid work? Everything we know suggests that nearly all people seek to make some contribution. And many of us believe that it would be positively awful to try to turn all socially useful contributions into waged employment. On this background, even the principle 'To each according to her

---

taxation as the basic source of funding. Nor do they avoid generating a net cost in terms of real disposable for some households and thereby raising an issue of 'affordability'.

contribution' justifies a modest UBI as part of its best feasible institutional implementation.

But a more fundamental reply is available. True, a UBI is undeserved good news for the idle surfer. But this good news is ethically indistinguishable from the undeserved luck that massively affects the present distribution of wealth, income, and leisure. Our race, gender, and citizenship, how educated and wealthy we are, how gifted in math and how fluent in English, how handsome and even how ambitious, are overwhelmingly a function of who our parents happened to be and of other equally arbitrary contingencies. Not even the most narcissistic self-made man could think that he fixed the parental dice in advance of entering this world. Such gifts of luck are unavoidable. And, if they are fairly distributed, they are unobjectionable. A minimum condition for a fair distribution is that everyone should be guaranteed a modest share of these undeserved gifts.[15] Nothing could achieve this more securely than a UBI.

Such a moral argument will not be sufficient in reshaping the politically possible. But it may well prove crucial. Without needing to deny the importance of work and the role of personal responsibility, it will save us from being over-impressed by a fashionable political rhetoric that justifies bending the least advantaged more firmly under the yoke. It will make us even more confident about the rightness of a UBI than about the rightness of universal suffrage. It will make us even more comfortable about everyone being entitled to an income, even the lazy, than about everyone being entitled to a vote, even the politically indifferent.

---

[15] Along the same lines, Herbert A. Simon counters the objection that a UBI would be unfair by observing 'that any causal analysis explaining why American GDP is about $25,000 per capita would show that at least 2/3 is due to the happy accident that the income recipient was born in the U.S.' He adds, 'I am not so naive as to believe that my 70% tax [required to fund a UBI of $8000 p.a. with a flat tax] is politically viable in the U.S. at present, but looking toward the future, it is none too soon to find answers to the arguments of those who think they have a solid moral right to retain all the wealth they "earn".' See Simon's letter to the organizers of BIEN's seventh congress in *Basic Income* 28 (Spring 1998).

CHAPTER 1

BASIC INCOME CONFRONTED WITH SOME POPULAR IDEAS OF JUSTICE

1. INTRODUCTION

Since the late 1970s, massive and longlasting unemployment is the primary problem for social-economic policy in European welfare states, and to a lesser extent, also in the US. Especially long spells of unemployment and the so-called 'modern poverty' are not only corrosive for the persons concerned but also for society at large. Governments try to attenuate the consequences of unemployment and poverty by providing social benefits conditionally, and, in so far as in its power, to take employment-promoting measures. However, recent social-economic policy measures designed to reduce unemployment can largely be characterized as 'piece-meal social engineering'. They vary from reducing the level of minimum wages, reducing the tax wedge, the introduction of work- and learnfare programmes for the unemployed, relaxation of firing and dismissal procedures, the implementation of work subsidies for low wage workers and more severe conditions for obtaining social benefits. All these measures are piece-meal because they keep the *conditional* nature of the arrangements of the welfare state intact. Almost all social benefits are in one way or an other connected with paid work: one is either too young or too old to do paid work (child allowances, state pensions), involuntary unemployed, disabled or sick (social assistance, unemployment and disability benefits) or preparing for work (scholarships). Job creation programmes and active labour market policies where the unemployed are assigned socially useful tasks to retain their benefit also strengthen the conditional nature of social benefits.

There is, however, an alternative available which is not the such-and-such adjustment but a major reform of the social security system. This alternative, an unconditional system of guaranteed minimum income, is known as basic income (henceforth BI) or negative income tax (NIT).[1] Unconditional stands here for the fact that a BI is paid irrespective of labour market history, present status on the labour market, willingness to work, wealth or income and household composition. A BI will be paid out to all citizens unconditionally, that is 'no questions asked'.

In light of the problems faced by the welfare state, it is useful to examine why the BI scheme has been proposed. Barry (1997) mentions at least four advantages of BI: it reduces dependency,[2] it boosts low wage employment[3] and part-time work,[4] it

---

[1] The equivalence between a BI and a NIT scheme is illustrated in chapter 4.
[2] 'Dependency - the dependency of a worker on an employer or a woman on a man – has rightly be seen [by socialists, LG] as the enemy to be overcome... it is not absurd to suggest that a subsistence-level basic income is a far more plausible institutional embodiment of it than anything Marx himself ever came up

ensures greater scope for compensatory justice (see chapter 2) and finally it stimulates co-operative enterprise.[5] The same advantages and some more, can be found in the following concise opening statement by Van Parijs (1992, 3): '... it [BI, LG] has been vindicated, using the widest range of arguments. Liberty and equality, efficiency and community, common ownership of the earth and equal sharing in the benefits of technical progress, the flexibility of the labour market and the dignity of the poor, the fight against unemployment and inhumane working conditions, against the desertification of the country side and interregional inequalities, the viability of co-operatives and the promotion of adult education, autonomy from bosses, husbands and bureaucrats.'

However, the idea of BI is still highly controversial because it hits the 'moral core' of the existent welfare state, which provides social benefits conditionally, temporary and selectively. Implementing a BI would mean a major change in course, a new orientation on questions of social security and equality of opportunity. Given that most welfare states have recently abolished exemptions from the obligation to work for certain groups (e.g., one-parent families and older, non-pensioned, workers) and are now moving more in the direction of an across-the-board obligation to work and towards work- and learnfare, eliminating all conditions on receiving minimum income would mean almost a U-turn in policy. Thus, the challenge the BI proposal poses is enormous. It not only has to be shown that it is economically feasible, but it also has to be morally acceptable as well as socially and culturally durable (i.e., serve the transition from a predominantly breadwinner family society towards (fiscal) individualization, accommodate new forms of flex work contracts and meet the more general requirements of social security and fair equality of opportunity).

On top of that, the equity arguments held by BI advocates[6] do not seem to be shared by the majority of people, nor by the professional economists with academic degrees. Table 1 shows the results of a survey by the Dutch Social and Cultural Planning Office (SCP) at a time when BI was seriously debated in Dutch politics. Two thirds of the population is against a (partial) BI. A majority of almost 60% is in favour of moving towards a scheme of learn- and workfare, which strengthens the

---

with' (*ibid.*, 165). For a thorough analysis of the link between Marx' realm of freedom and BI, see Van der Veen (1991) and Van der Veen and Van Parijs (1987).

[3] 'Within a regime that gave everybody enough to live on to begin with, even low earnings would make for a net addition to income and provide a margin above subsistence level. A job that was manifestly worthwhile, and seen by everybody as such, might well get takers at rates of pay that nobody under the current dispensation would afford to accept' (Barry 1997, 166). This effect is largely due to the removal of the poverty trap and minimum wage legislation (see also section 2).

[4] 'Part-time work might well prove especially attractive, and would not run into any of the difficulties thrown up by the existing benefit system. It is not necessary to second guess the details to see the potential for the revamping of work that is offered by basic income' (*ibid.*, 167). This conjecture is in line with the fact that at this moment part-time work is almost entirely done by dependent partners (housewives) not entitled to social benefits and thus not subject to the poverty trap and not bothered by complicated administrative procedures of the local social services department.

[5] See Howard (1998).

[6] See e.g. Van der Veen (1991), De Beer (1993), Van der Veen and Pels (1995) and Van Parijs (1995).

link between (paid) work and income.[7] The support for BI increased by only 4%-point between 1993 and 1995. Among the professional economists the support for BI is stronger,[8] but still insufficient. Klamer and Van Dalen (1996, 265) surveyed the opinions among a representative sample of Dutch economists. On the position 'The government should reform the social security system along the lines of a negative income tax (or basic income)', 19.4% of the professional economists agreed, 26.7% agreed with some reservations, 38.3% disagreed and the remaining 15.5% had no opinion.

Table 1. Support for workfare and basic income (in %) among a representative sample of the Dutch population (SCP (CV 1993, 1995), 1996, Table 5.6, 183)

	workfare		BI	
	1993	1995	1993	1995
Agree	59	58	19	23
Disagree	29	32	65	64
No preference	8	9	9	8
No answer	4	2	6	5
	100	100	100	100

According to the SCP-survey, the predominant reason for this meagre support is that a BI disconnects the link between work and income. The question then becomes to what extent a BI is contrary to common notions of justice related to the link between work and income which are presumably incorporated in the workfare scheme, but not in a BI scheme. Does a BI run against some strong and basic equity intuitions

---

[7] This figure is not unique for The Netherlands. According to Arneson (1990, 1130): 'Even if cash grants worked more efficiently to boost the utility of disadvantaged persons than provision of employment, a program of state-guaranteed employment might be uniquely palatable to voters in modern democracies. Opinion surveys routinely show that the majority of citizens harbor grave qualms about the wisdom of a state policy of handing out unearned income to the able-bodied but support programs that offer employment opportunities to able-bodied needy persons.'

[8] This is in line with what Burtless (1990, 68) says: '... it is safe to say that the negative income tax has had only one major constituency – economists. The public strongly rejects one key element of a NIT, the minimum income guarantee. Even politicians who have embraced the idea insist on adding features to the plan that are not part of the original conception, such as mandatory work requirements or guaranteed public jobs. Though a cash NIT has never been adopted, the debate over the NIT affected the terms of the welfare reform debate.'

held by the general public?[9] It goes without saying that a substantial BI reduces the need to be self-reliant (self-supporting) and allows parasitism. The parasitism objection states that a BI is in conflict with the notion of reciprocity. The unconditional nature of BI seems to run counter to the demand of 'if you do not work, you shall not eat' which is part of the neutral work ethic and does not fit in nicely with the slogan 'there is nobility in labour' belonging to the perfectionist work ethic. This is undeniable, but the conflict between these widely adhered to notions of justice and BI can be qualified. The rest of this chapter deals with the apparent conflict between BI and the notions of self-reliance, reciprocity and the work ethic.

These three popular notions are highly connected, but not interchangeable. Reciprocity may be characterized as doing or giving something in return for something that has been done or given to you. This mutual relationship is not inherent in the notion of self-reliance or the work ethic. Self-reliance is mainly the idea that one does not need to get financial help from the state in order to survive. The work ethic can simply be described, following Arneson (1990, 1127), as the idea that for the able-bodied persons there is an obligation to work, at least for those who are not wealthy or supported by someone else. This obligation to work need not solely be motivated by the requirement to be or become self-reliant. For instance, a workfare programme offering jobs to do socially useful services without improving the skills of the workers is in line with the work ethic (and the demands of reciprocity), but not with the aim to increase self-reliance. Similarly, all kinds of conditions, duties and obligations which have to be met in order to become or remain entitled to social benefits may satisfy the demands of reciprocity, but need not improve the self-reliance of the benefit recipients.

In the following three sections, the apparent conflict between BI and self-reliance, reciprocity and the work ethic will be explored. At the end of the next section on self-reliance, a few remarks are made concerning workfare and wage subsidy policies to improve self-reliance. In section 3 the reciprocity-based parasitism objection against BI will be discussed. Section 4 draws a distinction between a perfectionist and a neutral work ethic. An overall evaluation of whether the idea of BI can endure the challenges posed by these three popular notions of justice concludes the chapter.

2. SELF-RELIANCE

Self-reliance is an ambiguous concept. According to Van Heerikhuizen (1997, 184), from a sociological point of view *individual* self-reliance does not exist because people can only survive in communities. The human condition is one in which dependency on others creates reliance on them as well. Using this concept, self-reliance may only be discussed in relative terms by summing up all various

---

[9] There is a substantial amount of literature about the role of need, desert, contribution and equality in a just income distribution according to the opinion of the general public. Unfortunately, this type of research never specifically focused on the role of a BI in this respect.

interdependencies in the social fabric. This narrow view acknowledges that even someone of independent means is still dependent in numerous ways (ranging from the deliverance of food in supermarkets to the protection of private property by the state). This view does not paint a complete picture however. The common sense understanding of self-reliance incorporates the view that each citizen has a responsibility for securing their own livelihood or 'earning her keep' (at least for those who are not supported by their spouse, family or friends and who are not wealthy).

At first glance, a BI may appear contrary to the common sense understanding of self-reliance. Ideally, a social security system based on self-reliance will help the poor to help themselves, to attain and retain the capacity for self-care and self-support and thus eliminate dependency. A BI, however, may enable or foster massive dependency by shifting the burden of the cost of livelihood on to the state. A BI legitimizes this dependency on the state by granting a BI as a matter of right. Furthermore, while the popularity of the self-reliance notion and related notions (e.g., responsibility, independence, self-respect, self-esteem and self-confidence) among right-wing politicians is partly due to the belief that social welfare programmes may do more harm than good,[10] the real objection with social welfare programmes is not in that people now and then receive help, but that people begin to *rely on it* (its legal entitlement, as a matter of right).[11] The argument also insists that social assistance should be given only in case of need or emergency in such a way that recipients do not become dependent on the help of others or the state (and so not changing their behaviour). Now a BI is exactly the opposite, because it is given to all as a matter of right, whether needed or not and on a permanent basis. Admittedly, a BI removes the *necessity* to be self-reliant (self-supporting), but as we will later see, it also makes it easier for those for whom self-reliance is now a problem to find paid work and therefore reduce dependency.

Before examining whether or not a BI violates the popular idea of self-reliance, it is necessary to investigate what the principle itself is worth. According to Goodin (1988), there is a serious flaw in the argument for self-reliance. The main flaw in the right-wing politicians' advocacy of self-reliance is revealed in the implementation. Self-reliance can be laid down in social legislation by means of more severe conditions for receiving income support. This is done by extending the means-test

---

[10] See for instance the main question posed by Murray (1984, 16) 'How is a civilised society to take care of the deserving without encouraging them to become undeserving?', and his principles for designing social welfare programs that 'Any objective rule that defines eligibility for a social transfer program will irrationally exclude some persons' (*ibid.*, 211), and, 'Any social transfer increases the net value of being in the condition that prompted the transfer' (*ibid.*, 212). To be sure, these two problems are more relevant for extensively targeted conditional schemes than for a BI scheme, due to its unconditional nature.

[11] The right-wing plea for self-reliance or economic independence is strongly connected with the principle of self-ownership and voluntarism. Given the distribution of property, each person is the owner of his talents and the fruits it will bear on the market. The market is based on voluntary exchange, so those who cannot help themselves are dependent on voluntary gift-giving by others. A voluntary gift does not violate self-ownership and it makes it difficult for those who receive it to rely on it (the gift cannot be enforced, at most expected). The strong version of self-ownership forbids any kind of compulsory redistribution of the rich to the poor (see Cohen 1995), whereas the weak version is not at all in conflict with a BI scheme (see Van Parijs 1995, chapter 1).

from the core family (usually the partner only) to the extended family (parents, grandparents, grown-up children). This is necessary to ensure that income transfers from wealthy family members to needy ones can be enforced. However, this runs counter to the principle of self-reliance in the same way as when the needy have a legal claim to a transfer from the government. As Goodin (*ibid.*, 343) rightly points out: 'Surely no one is *self*-reliant when he has to rely for support on others, whether they be family, friends, or state officials. The only way to make any sense at all of this otherwise perplexing notion is to understand that, for purposes of this argument, the boundaries of the "self" have been extended to include one's household (or extended family, or social network) as a whole. Whatever its political appeal, such an expanded notion of the self is philosophically preposterous.' Moreover, aside from the problem that the incidence of poverty is particularly strong among poor families (no family-member is in a position to monetarily support others), it brings forward other problems: 'Family ties cannot be maintained or strengthened by statutory enactments.... Litigation ... is often very painful to those in need of help, ... does not yield any returns in family solidarity, and ... yields monetary returns which are far below the expense of litigation' (Abbott, cited by Goodin (1988, 347)).

   The alternative is that welfare state provisions are limited and income support is left over to *voluntary* support given by relatives, friends and charity organizations. This alternative of shifting the burden of poverty relief to voluntary actions is internally contradictory because then '... self-reliance asks two contradictory things of people: on the one hand, "prudential regard to their own future"; and on the other, "effacement of self in response to claims of helpless relatives"' (*ibid.*, 348). The contradiction is particularly relevant with respect to women. It is questionable whether the family as a first line of defence is favourable for the self-reliance and independence of women. If public health, child- and elder-care are curtailed, a much greater burden is placed on women to take over these tasks. They can only do so by forgoing the opportunity to take (full-time) paid employment, giving up paid employment altogether, or worsening their prospects of future paid employment, which may all compromise their economic independence. Therefore, reducing or eliminating state dependency only transfers the dependency to others in so far it does not alter the amount of dependency, but state dependency has the advantage of being non-discretionary and alters its quality. Contrary to voluntary help, it is not charity, but a right laid down in social legislation.

Why then is the notion of self-reliance still so popular. The reason may be this:

> Self-reliance is valued most highly as a character trait... This emphasis on character is also betrayed by our choice of *which* dependencies we find offensive. 'When we think of... dependency..., we neglect the student preparing for a socially useful occupation, the mother caring for children, the aged who have earned retirement, the workers who are forced out of the labor force by technological change... [Instead] our attention is rigidly fixed on those who should be working and are not,' and whose dependency therefore would seem or betray a flaw in character (Rein, cited by Goodin 1988, 354).

On this account, dependency is objectionable only when those concerned could have avoided it on any reasonable terms. Surprisingly, the BI proposal is, to say the least, ambiguous in this respect. In one respect, a BI makes it possible for all citizens to rely on long-term minimum income support, and for some 'nonneedy bohemians' this may be sufficiently attractive to abstain from any (search for) paid employment. In another respect, the elimination of the poverty trap and the creation of employment opportunities at low wage rates not existent in present welfare states would give those at the bottom of the labour market more incentives and more chances to find paid work. Better incentives arise because of the elimination of the poverty trap. Social security recipients face now a very high withdrawal rate, because not only all net labour income up to the social security benefit is withdrawn, but also because subsidies for other expenditures are reduced.[12] Moreover, working may engender additional costs (e.g., travelling costs to work are only partly compensated). On the other hand, a job may entail fringe benefits (ranging from firm trips to employer paid private telephone calls) and more importantly, the opportunity to improve and expand one's skills and to make promotion. In practice, however, for those with the least marketable assets it will be difficult to escape being trapped in poverty, at least in the short run.

With a non-means tested BI, all gross labour income earned, no matter how small, adds to total net income. Removal of the poverty trap improves the incentives for the unemployed, but does little for their chances to find employment. The level of the social assistance benefit may be the most important determinant of the reservation wage level for the low-skilled unemployed under a conditional scheme, especially since the social benefit is lost when they resume employment. Moreover, most welfare states have minimum wages. Under a BI scheme, there is no need to have minimum wage legislation in order to protect workers since all have free access to a real alternative, to live from a BI only. Because the level of the BI does not act as a *direct* determinant of reservation wages (the BI is received irrespective of labour market status), potential workers have only to balance and compare the sacrifice they bring in doing a job (e.g., giving up leisure time) and the (non)pecuniary rewards of a job. Therefore, it is the type of job and the effort and disutility of doing that job that determines reservation wages. Depending on demand and supply conditions under a BI scheme, there may be a large number of jobs where gross wages are below current gross minimum wage levels (and so at the moment not undertaken, or located in the black market). Removal of the minimum wage constraint may thus increase the chances for the low-skilled unemployed to find work, if they want to.[13] In addition, under the present scheme the unemployed may prefer the security of the social benefit above the insecurity of a temporary job, just because they are not certain whether they will be entitled to the social benefit once the job is lost. In short, although a BI would introduce an unconditional minimum income to be received from the state, and thus institutionalizes

---

[12] The elimination of the poverty trap under a BI scheme is of no use for those who receive several earmarked income-dependent subsidies at the same time. The effective tax rate (inclusive the flat earnings tax rate of the BI scheme) may well be near 100%. To solve the problem of the cumulation of (implicit) tax rates, one can think of an anti-cumulation measure to limit the effective, cumulated, tax rate significantly below 100%.

[13] The use of tax money to offer all kinds of jobs programmes to the involuntary unemployed can also be reduced under a BI scheme since they can use the BI as a kind of wage subsidy to price themselves into a job, at least if their productivity is non-negative.

dependency for those who pay less in taxes to the BI fund than they receive in the form of a BI, it also makes it easier and financially more attractive for those who are now on welfare (and thus not self-reliant) to find paid work and reduce their dependency.

The proposition that BI reduces self-reliance is therefore not a conclusion that follows logically: it is motivated on empirical and moral grounds. The empirical issue hinges on how many people under a BI regime wilfully and intentionally choose to live from the BI only and forego productive use of capacities and talents in paid employment. In any case, the danger of large-scale parasitism will be smaller at lower levels of the BI. The occurrence of any wilful parasitism will, in addition to the level of the BI, depend on the distribution of potential earning powers and on preferences for work and leisure. The higher the potential earning power from using one's talents, the greater the opportunity cost of abstaining from paid employment. Also, the higher one's value of leisure time, the greater the opportunity cost of doing paid work. Finally, the stronger one's work ethic and the more attractive one's work, the more likely it is that one will remain employed, even if net wage rates are lower and a BI constitutes a significant portion of total disposable income.

Taken together, the following effects may be expected. The labour supply of high income earners will probably not change, because they are already subject to a high marginal tax rate (a tax rate comparable to the flat rate required to finance a BI). Because the introduction of a BI would reduce their total net income, it may create a positive labour supply response through the income effect. The labour supply of low and middle income earners will probably be reduced, because the tax rate needed to finance the BI is higher than the marginal tax rates they are facing now and because the BI raises total net income if they do not change their labour supply. For these two groups, of whom as a rule the males work full-time, it will become more attractive to reduce their labour supply, because compared to the existent scheme a reduction in the number of hours worked will have a smaller effect on total net income under a BI scheme.[14]

A BI is most favourable to the disadvantaged with a low potential earning power, who are unemployed but prefer to do paid work. A BI system makes it possible for them to earn as much money they can and to become as self-reliant as possible.[15] It may also help them find jobs, not only because there are more job opportunities at wages below the present minimum wage levels, but also because others in the lower and middle income groups may have reduced their labour supply.[16] For those who

---

[14] For a more detailed examination of these effects distinguished by groups of workers, see chapter 4.

[15] There is one aspect of current welfare state programmes that is especially detrimental to the plea of self-reliance for those who once in a while have to rely on social assistance, namely that most social assistance programmes are means-tested. Those who have saved some money in the past are punished more severely than those who did not save. A BI does not have a means-test, so it does not impose a penalty on past prudence. From this respect, an increase in savings, and thus a stimulus to self-reliance, among those groups is expected.

[16] If that would indeed be the case, the introduction of a BI would probably have the effect that labour supply measured in the average number of hours per worker decreases, but that the number of (part- and full-time) workers increases. The combined effect may reduce or increase the total labour supply, but it would certainly mean a greater dispersion of paid work among the labour force.

are responsible for caring for children or other relatives or do the housekeeping, it will become more attractive to abstain from paid work (of course this should not be qualified as a morally objectionable kind of parasitism on the greater capacity for self-reliance of others).

In sum, for the main categories of the labour force, the picture of self-reliance does not look so bad. Although middle income earners may reduce their labour supply, this does not mean that they reduce their degree of self-reliance, at least as long as they pay more taxes to the BI fund than they receive out of it. These wage earners just prefer to substitute in more leisure. The most negative effect in terms of self-reliance has to be expected among the low-income earners, especially if many of them decide to reduce their labour supply significantly.[17] For the disadvantaged, a BI may offer more opportunities to increase their degree of self-reliance, but it also offers the opportunity to forego any progress towards self-reliance. Indeed, opponents of a BI may feel that the extent to which able-bodied persons will be self-reliant is higher when the negative income consequences of not being self-reliant are higher. Broadly conceived parasitism must be expected to be concentrated among those with low earning power and a high value of leisure time. A large part of this group is made up of women taking the responsibility of household tasks. As we have seen above, non-self-reliance among this group cannot be attributed to 'a flaw in character'. Narrowly conceived parasitism is limited to those who wilfully and intentionally choose to abstain from paid work because of laziness or the desire to free wheel among non-needy bohemians. This last phenomenon, the layabouts and good-for-nothing, is the real moral issue. This issue is taken up in section 3, which deals with the objection that a BI violates a widespread norm on reciprocity.

To say that a BI scheme is not necessarily in conflict with the demand of self-reliance does not imply that there are no other alternatives that are more conducive to the pursuit of this ideal. In particular, one can think of a very harsh regime in this respect. However, there is no broad support for a return to a kind of poor laws regime like those experienced in the nineteenth century. Thus it is necessary to compare the present welfare state, a residual oriented welfare state,[18] a workfare-wage subsidy scheme and a BI scheme, to determine which is most favourable with regard to the principle of self-reliance. A workfare-wage subsidy system is more in line with the principle of self-reliance (that is, to help people to help themselves) than the first two systems. The amount of money the first two systems spend on active labour market policy through job training and schooling programmes for

---

[17] Goodin (1988, 339) is rather optimistic in this respect when he writes: '... the programs upon which recipients might be more inclined to rely are the ones most like genuine insurance programs (e.g., retirement, unemployment insurance, disability insurance, workmen's compensation); social assistance programs targeted more tightly on the poor generally seem to have much weaker incentive effects upon recipients. Similar findings emerge from experiments with a negative income tax. Overall, beneficiaries seem to reduce labor supply only slightly (by, perhaps, five percent), while those in the very poorest households sometimes actually increase it. In short, the poor – and the poorest of the poor, most especially – tend not, for the most part, to work less hard when they know that they can rely on state support. That is, if anything, a vice of the middle class...'

[18] With residual oriented welfare, the state only provides a conditional minimum income guarantee and leaves all other (above the minimum income) arrangements to the private insurance market.

unemployed is rather small compared to the expenditures on social assistance in the form of income support. The crucial test is therefore between the workfare-wage subsidy and the BI scheme. Both have the advantage over the first two schemes in that they not only relieve poverty by means of income transfers, but also try to fight unemployment.[19]

In a workfare-wage subsidy regime, the unemployment benefits are conditional upon the readiness of claimants to perform unpaid but socially useful services, and employers are offered wage subsidies for employing formerly unemployed. Assigning useful social services to the unemployed gives government officials and all those organizations who are allowed to offer projects much discretion in determining what is socially useful; under a BI scheme, this is determined by the forces on the cultural market place. This is certainly something that should be welcomed by libertarians.[20] There is also a serious danger of crowding-out formerly paid employment, the more so if involuntary unemployment is high. Even more important is that workfare measures do not directly enhance self-reliance,[21] at least not in the short run: it is only a way to prevent people on the dole from getting social security benefits without doing something in return.

Wage subsidies[22] are a more efficient and direct way to enhance self-reliance, but they too have a few major disadvantages. The danger of crowding-out regular jobs is even greater for wage subsidies paid to employers than for workfare programmes organized by the government. Further, part of the attractiveness of paid work is the social recognition of its usefulness and the power it confers on the job holder,[23] and this is destroyed by wage subsidies. The amount of subsidy required to create these jobs reflects the extent to which the cost of the activity outweighs the (short term) benefits. How does a BI system behaves in this respect? It was argued earlier that all workers with a low hourly gross wage rate are implicitly subsidized (they are net recipients under a BI scheme) by those with high incomes. Can it be concluded from this that under a BI system the jobs occupied by low wage earners

---

[19] Moreover, the more successful the first two systems are in guaranteeing a means-tested minimum income, the greater will be involuntary unemployment (and involuntary employment) since higher social benefits increase efficiency wages above market clearing wages (which has a detrimental effect on the 'equilibrium rate of unemployment' (see Van Parijs 1987, 1)) and the more severe the poverty trap.

[20] Another problem is that a government that acts as 'employer of last resort' (a looming phenomenon if one really wants to make workfare work) may lead to a 'big government', something what libertarians usually do not like.

[21] See also Plant (1993, 46): 'It is argued that self-respect comes from getting a job and sticking with it. However, if a job is created by government or a government-funded agency for an unemployed person, is this likely to create a sense of personal satisfaction and self-respect for that person? It strikes me that there is all the difference in the world between arguing that self-respect comes from getting a job in the labour market and having a government-funded job created for you... it is transferring dependency from the Department of Social Security to the Department of Employment.'

[22] E.g., the much discussed voucher proposal of Snower (1994), where long term unemployed can use part of their social benefit as a wage subsidy, the level of which depends positively on the spell of unemployment.

[23] '... having a paid job constitutes a social recognition of one's usefulness, of one's worth. It shows that society cares for what the individual has to offer, for otherwise it would not pay for what he does' and '... a paid job also bestows upon its incumbent some power over his employer and over society at large. The very fact that someone is willing to pay for his work shows that what he produces is worth something to others, and hence that refusing to produce it is a cost to society' (Van Parijs 1987, 2).

or activities initiated by self-employed which yield low profits, are subject to the same loss of social recognition of usefulness and power? The answer is partly negative, because all workers and self-employed are subject to the same scheme, since they all receive a BI which can be used in the way each citizen chooses:

> ... it is up to the individual beneficiary to decide whether (s)he will turn it into a straight wage subsidy, by accepting to do the job (s)he is offered at a substantially lower cost to her employer. (S)he may also use it instead to create her own job – whether as a self-employed individual or as part of a partnership or cooperative – or even convert it into a subsidy to unpaid activities, whether productive or unproductive (Van Parijs 1987, 5).

The crucial difference is that wage subsidies (and workfare programmes) are targeted to those who have proved themselves unable to find work on prevailing terms, while a BI is given to all irrespective of their past (lack of) labour market success. However, all those who pay taxes to the BI fund which fall short of the amount of BI received are implicitly subsidized by others for whom it is the other way around.

In conclusion, a BI creates a situation in which the necessity to be or become self-reliant is weakened or even absent, but this does not automatically mean that the degree of self-reliance across the labour force declines. Instead, the degree of self-reliance is highly dependent on the distribution of preferences. If preferences (or the work ethic) are such that a decent BI is economically unfeasible, then there is even no point in arguing for a BI. If the work-ethic is sufficiently strong, then there are good reasons to believe that most workers will remain self-reliant and that those who are now unemployed can increase their degree of self-reliance. As will be seen in the next section, there is even a more serious objection: even if opponents may grant that a decent BI is feasible, they may still object to the freewheeling lifestyle of artists and others.

3. RECIPROCITY: NOT ONLY THE TRULY LAZY

Opponents of a BI may fear that the institution of a BI does not only give way to laziness and inactivity, but also invites freewheeling by the so-called nonneedy bohemians. There is no guarantee that recipients of a BI who do not perform paid labour will volunteer work, take care of family members, or perform other socially useful activities. They certainly use up part of society's scarce resources, and some may do nothing in return, or do something in which nobody is interested. Here we touch upon a strong and widely adhered to intuition, namely reciprocity. In a social security system based on conditionality the government has at least the possibility to ask reciprocal services in return for the benefit, as is the case with a workfare system. The right to a social benefit can thus be balanced by a duty to do socially useful services, or by a duty to resume working as soon as possible. However, due to the unconditionality of the BI, the government can do nothing against do-nothings, lazybones and spongers or against the freewheeling surfers and artists. The parasites cannot reply that this no-work option is also available to others, because it is just the moral acceptability of the option which is the issue. This reciprocity-based parasitism objection is a ponderous argument against BI. Aside from the apparently blatant violation of the work ethic, the really interesting question is whether laziness and freewheeling on a large scale is a legitimate threat. First I will give some qualifying remarks concerning the position of the truly lazy and then turn the attention to the freewheelers.

It cannot be excluded that some will adopt a conception of the good life which qualifies them as truly lazy, but is it plausible that a large part of the population will do so if they can? Levine (1995) argues that the demands of reciprocity and equal respect (neutrality to different conceptions of the good life) are mutually conflicting, and that as affluence increases, the urgency of reciprocity recedes in favour of a state-supported right to be idle. I do not want to delve further in this intricate argument in favour of the lazy, but only wish to add that according to Levine '... most genuinely autonomous agents would find productive activity (pursued in or out of the monetized economy) an intrinsic part of their conceptions of the good, and because under real world conditions productive activity is generally attached to paying jobs, I am not worried that, in a genuinely free society, many people would choose unemployment or significant underemployment regardless of how a right not to work is implemented. I therefore doubt that a right not to work will prove unacceptably costly' (*ibid.*, 265). If this optimism is shared, then the real danger of a BI is probably not large scale *inactivity*, but activity *uncontrolled* by either the state or the market. This lack of control on those who choose not to perform paid work, although able to do so, is indeed a feature of the BI proposal which can be positively or negatively evaluated. It is not fanciful to believe that under a BI scheme a strong proliferation will occur of non-marketable activities. These non-marketable activities may show a great diversity, ranging from joining Greenpeace to desperately trying to make a living as a musician.

With a substantial BI sufficient to provide for basic needs, everyone is free to do as he or she likes, irrespective of whether there is a market for it or not. Seen from this perspective, isn't a BI a license to engage in activities without any concern for societal needs and wants? Isn't it the road to abolish the market as a disciplining device on one's activities? The provision of a BI gives artists, e.g. all those who have studied unmarketable professions like arts, philosophy, music, etc., the opportunity to do only what they find interesting. Admittedly, the set of choices (or lifestyles) open to everybody is greater under a BI than under a conditional scheme: everybody has the real choice whether to work or to indulge in leisure spent in an artistic way. However, advocates of conditional social security will object that fortunately there is no such thing as a state supported right to freewheel, and that maintaining the link between work and income has priority above expanding choice sets. Although the choice set open to the artists may be widened by introducing a BI, it nevertheless is not wise and right to introduce it because the cost of providing this extra choice to those who are not performing paid work is in the end paid for by those who perform paid work (their opportunity set is reduced since they will receive less net income for the same work effort). There is no right to an interesting job, nor a right to do only those things one finds interesting. In general, wages received by workers reflect the value others attach to their labour. These wages depend on the amount of marketable and hence exchangeable assets which each worker has, and how eager they are to use these assets productively. In the market place, one is not paid for the value of possessing these assets and their use for the holder or producer, but for the value they hold for the purchasers. The market serves as a disciplining device making agents keen to direct and shape their activities and, perhaps more important, their skills with an eye to the value it offers for others.

In summary, under a conditional scheme the activities and interests of some citizens (here for the ease of argument called artists) is not seen as a legitimate ground for receiving benefits. It goes too far to argue for a BI just to accommodate artists' wishes, but it is clear that under a BI scheme they are in the position to follow their preferences. In the same vein, it cannot be ruled out that under a BI scheme many more students will chose the kind of educations with a high leisure-time value and a low market-value. The other more positive side is that not only the set of choices of non-workers expands, but also that of workers: for all workers, especially low wage workers, there is more scope to search only for what they see as a good job. The possibility to abstain from any form of paid employment if jobs available to them do not offer what they are looking for, may serve as a kind of disciplinary device for employers to improve working conditions. However, a BI system is not a system which offers everyone a right to an interesting job, let alone to guarantee a job for all those who seek work. What a BI does in this respect is that it gives to all the real opportunity to balance the net utility of a job offer to the net utility of leisure (including a BI). If workers find that the first side of the balance is in the red, either the reward for the job must be increased or its conditions of employment must be improved. This theme, that a substantial BI can serve the interests of low-skilled workers for greater compensatory justice, is taken up in chapter 2. There I argue that there is a tradeoff between the endeavour to reach a state of affairs with no-parasitism and a state of affairs with compensatory justice.

So far the attempt to bring the idea of BI in line with the demands of reciprocity is not exactly a success story. However, as the example of the artists illustrates, an inherent difficulty in using the parasitism- or reciprocity-argument against BI is the lack of a clear-cut criterion for assessing the social usefulness of all kinds of activities (this is a problem which ideally need to be solved under a workfare scheme, but not under a BI scheme). Admittedly, the alleged social uselessness of freewheeling activities and the casualness with which some may take the liberty to freewheel under a BI scheme may go against the grain of the general public, but the reciprocity argument is not suitable in this respect. To use the reciprocity argument here would mean that just because of the lack of a clear-cut criterion to evaluate social usefulness, it is legitimate to use all kinds of disciplinary measures to ensure that everyone does the things which are considered socially useful according to some objective standard or consensus. There are two further considerations. Firstly, when comparing the present and the BI social security scheme with respect to whether the demands of reciprocity are really met, one has to compare the number of persons who receive the BI but do not 'deserve' it (because they do not make any reciprocal contribution to society) with not only the number of persons who are now 'non-deserving' social benefit recipients but also with all 'deserving' recipients who are locked in a position that they cannot make a contribution (e.g., because of the rule that it is not allowed to enrol in formal schooling or doing full-time volunteer work while on welfare). If the group of 'non-deserving' BI-recipients is small, then one has to think about whether we are prepared to put up with these 'malafide' recipients of BI. This brings us to the second consideration, namely the disbursement of BI to all, subject to some reasonable and broad reciprocity-test (a list containing a wide range of approved activities including care work, training, sabbaticals, extra leaves to recover from burn-out, etc.). This proposal, known as *participation income*, was made by Atkinson (1993; 1995b, chapter 15; 1996). Except for the reciprocity-test it shares with the BI that it is individualized, not means-tested and independent of past and present labour market status. The main advantage is that this system allows a wide range of unpaid, but contributive activities (and thus excluding the more extreme freewheeling activities) instead of the narrow focus on paid work and government-organized socially useful workfare activities. Van der Veen (1998, 159) believes that such a participation income scheme is ridden with insurmountable problems:

> ... there will be huge disagreement as to what this comfortable-sounding way of framing the reciprocity principle should include, both with respect to the list of 'socially useful activities',... as well as the duration and quality of performing selected items on the list (should negligently intermittent care work pass?). Whatever the consensus on these matters, the most pressing problem for a reciprocity-minded policy of participation income will be how to monitor and sanction the variety of multi-interpretable behaviour, and what powers of discretion to give to the administrative officers involved in the process ('Good morning. You mind if we go through last month's update of your contributions to worthy causes?').

Given these problems and provided the number of truly 'non-deserving' recipients of BI is low, it might be better and more efficient to take the freewheelers and lazybones for granted and remit a BI unconditionally.

4. BASIC INCOME AND THE WORK ETHIC

Many documents issued by government bodies and political parties preach the work ethic, and not surprisingly. The duty to work is one of the corner stones of present social security, becoming manifest in particular under workfare. Most social security benefits are in one way or the other tied to paid work. Advocates of the work ethic may either use a *perfectionist* or a *neutral* stance.[24] I will be brief about the former and concentrates on the much stronger conflict of the latter with the idea of BI.

According to the perfectionist point of view, paid work has intrinsic values for the worker and for society. Because paid work means earning an income, developing one's skills, structuring one's time, having social contacts, enforcing one's self-respect and so on, it is seen as the most important constitutive way of personal development and personal responsibility. For society at large, solidarity (understood as the willingness to pay taxes to finance social security) and integration (e.g., immigrants) rests for an important part on active participation in the labour process.

---

[24] For this distinction, see also Heij *et al.* (1993, chapter 11).

The problem with the perfectionist stance is that even if we agree with the characteristics it ascribes to paid work, it does not imply that a liberal government must force its citizens to do paid work, as the perfectionist stance recommends. For instance, to have children may also structure one's time, develop one's skills, provide social contacts and enhance one's self-respect. By itself this is not a sufficient reason to enforce a rule that all people who are able to have or raise children must do so. Moreover, if paid work has such a high intrinsic worth, are adherents of the perfectionist view not obliged to advocate paid work for all women too? Isn't it true that unpaid work has to a large extent the same characteristics? To impose a duty to work due to its integrating capacity, is also hardly defensible. Even if we grant that paid work is an integrating device, it does not follow that the government is entitled to impose it on work-shy citizens, for the simple reason that many more activities can be considered integrating devices, where it would not be just to make them compulsory.

The neutral work ethic does not take into account the alleged intrinsic values and takes for granted that some people dislike paid work, or think that it will not make a valuable contribution to their well-being. Nevertheless, defenders of the neutral work ethic maintain that the work-shy must bear the consequences of their voluntary choices: if you do not work, you shall not eat. Social security, so it seems, is compatible with a duty to work. It also is legitimate, following a neutral work ethic, to cut-back social benefits if recipients are unwilling to accept job offers, or to transform present social security arrangements into a workfare system. In sum, the neutral work ethic demands a conditional social security system and is therefore incompatible with the unconditional nature of the BI system.

One way to escape the conclusions of the neutral work ethic is to criticize its narrow perception of work. Not all work is paid work. Even more strongly, the number of hours spent on activities classified as unpaid work is higher than of paid work. It is also widely known that paid and unpaid work is not evenly distributed between the sexes. One advantage of a BI is that it can be interpreted as an implicit wage or reward (and thus also as a kind of social recognition) for the value of these activities, namely for those who carry the burden of keeping the household, raising children, caring for grandparents and grandchildren, and so on. Moreover, the argument goes that it is undesirable that these activities are transformed into the format of paid labour, with contracts, monitoring, quality control, time clocks, and the like.

For several reasons I do not think this is an effective way to counter the conclusions following from the neutral work ethic. For convenience, it may be useful to distinguish unpaid work into three categories: keeping a household, taking care of others and performing volunteer work. For keeping a household it is quite natural to be unpaid because the reward follows immediately: a clean house, or washed clothes. Usually this work is only paid for if it is done for others, i.e., if others buy these (household) services. How the burden of keeping the household is distributed between the household or family members is in the end an internal affair, or, if one member does paid work and another unpaid work, a kind of division of labour. Taking care of others, like raising children, clearly is a 'service' provided for others, but it would be quite odd to demand for it to be paid by them. It would also be unreasonable to ask society at large to pay for it, as the interpretation of a BI as a kind of implicit wage suggests. If children generate positive external effects, or if it is considered unjust that the decision to take children is too dependent on lack of income, there is more reason for the institution of child benefits (or even a parenthood benefit, or paid parental leaves), but not for a BI. Finally, even if we recognize the bias in stressing the importance of paid work at the expense of voluntary unpaid work for society, it does not lead towards a BI, but to a less austere policy concerning exemptions from applying for jobs for those who perform volunteer work, while they retain their benefit.

To take stock of the arguments given so far, the perfectionist stance towards work, expressed in slogans like 'there is nobility in labour' or in the policy of 'work above income', can be rebutted by invoking the same positive attributes to other activities. This shows that the argument is not strong enough to justify the duty to do or seek paid work. The objection against BI from the position of the neutral stand towards work, expressed in the slogan 'those who do not work, shall not eat', cannot be countered by criticizing the narrow definition of work as paid work only, and by using instead a broader definition of work which includes unpaid work. The neutral work ethic, in contrast to the perfectionist work ethic not making use of intrinsic, integrating or solidaristic characteristics of paid work, can be interpreted as the demand that those who voluntarily choose not to do paid work should bear the full cost of their own choices.[25] However, as will be argued in the next two chapters, a BI can be seen as a compensation for giving up one's fair share of job rights under unemployment, and parasitism as a manifestation of a breach of the neutral work ethic must be traded-off against the gain in compensatory justice.

SUMMARY AND CONCLUSIONS

The difficulty of returning to (near) full employment and especially the high incidence of unemployment at the bottom side of the labour market during economic slums has inspired economists to make various policy proposals. Almost all measures actually applied maintain or even intensify the conditional nature of social security arrangements. The implementation of a substantial BI would engender a major shift in social and economic policy. Even if the BI proposal proves to be economically feasible and sustainable, it would nevertheless not be a serious alternative to present social security systems as long as some popular objections against the proposal cannot be refuted or attenuated. In this chapter, the BI proposal has been confronted with the demands of self-reliance (self-maintenance, self-support, independence), reciprocity and the perfectionist and neutral work ethic. It is undeniable that the necessity of self-reliance or self-support is reduced as soon as everyone is granted a BI. This, however, does not automatically imply that people will behave in such a way that their extent of independence declines, certainly not when both the opportunities to and perspectives for those with the least marketable skills improve. Those who now are supported by means of social assistance benefits

---

[25] I thereby abstract from one reading of the neutral work ethic which says that those who have enough means to their disposal for a decent living (the wealthy), still have a duty to do paid work.

may have greater incentives and chances to become self-reliant under a BI scheme. By and large this effect is due to the elimination of minimum wages and the poverty trap under BI social security. In this respect it is important to note that the BI can be seen as a non-stigmatizing subsidy for all those with low earning capacities, while at the same it removes the 'bite' of both the poverty trap and the minimum wage legislation. It can also not be denied that under a BI scheme the government must abandon the conditions which are now attached to being a social benefit recipient. Therefore, there is no guarantee that recipients of a BI will reciprocate this gift of society with a countergift. As in the case of self-reliance, what one tries to exact from citizens (reciprocity) is here replaced by voluntariness. What about the work ethic, either understood in the perfectionist or the neutral variant? The perfectionist work ethic is rejected because it cannot be taken for granted that the beneficial characteristics ascribed to paid work are indeed unique for paid work alone and do not also apply to unpaid work or other activities (raising children for instance). It therefore cannot legitimizes an across-the-board duty to do paid work on all citizens. Moreover, it would also require that doing paid work is made obligatory to all women who are currently doing unpaid work. The duty to work imposed on social benefit recipients has more to do with reciprocity and the neutral work ethic. In the next two chapters I will present two arguments which might justify the breach of the neutral work ethic. To allow some parasitism can be seen as the price to be paid to attain compensatory justice on the labour market (chapter 2). In addition, if there is structural unemployment, the choice for the non-work option can be interpreted as giving up one's fair share of job rights in return for a BI (chapter 3).

CHAPTER 2

COMPENSATORY JUSTICE AND BASIC INCOME [1]

1. INTRODUCTION

What is the most plausible and feasible conception of compensatory justice (from now on CJ)? What conditions are required to achieve it? The general assertion of this chapter is that the more favourable the conditions under which people can freely choose their labour market careers, the more scope there is for CJ. More specifically, an economy with an unconditional basic income (BI) – one sufficient to cover the basic needs according to the prevailing standard of living – fulfils the conditions required from the (economist's) viewpoint of CJ, and is better than a conditional system of social security with a guaranteed social minimum at the same or even somewhat higher level. This topic is of interest for welfare state reform proposals in both Europe (characterised by high unemployment, modest income inequality and generous social benefits) and the US (low unemployment, high and increasing income inequality along with the rise of the working poor). If CJ is considered as an important ideal, then the present move from lax welfare to workfare (see Figure 1 in the Introduction) in both Europe and the US is not the right way. In the same vein, to top up the income of the working poor by means of an EITC as is the case in the US, or to give employers marginal wage subsidies to raise the demand for low wage workers, does not improve the bargaining position of workers towards employers, and hence does not help to achieve CJ.

Sections 2 to 4 provide a brief outline of three different approaches to CJ, one economic and two philosophical. I argue that the economist's view is the most fruitful, but still incomplete, and that the two philosophical (the objective and the balancing) approaches are impracticable. The discussion on CJ is linked with the more general debate on social justice by means of mapping the relation between the favoured conception of CJ to the Rawlsian maximin criterion (Rawls 1971). In the appendix, I show that a BI is an attractive alternative to implement the principle of fair equality of opportunity and to provide all a decent fall-back as part of Rawlsian justice. This is followed in section 5 by an enumeration of the more general conditions of CJ which must be met so that the market forces can do their proper work. Briefly, these conditions are: free career choices, absence of *in*voluntary unemployment, and a readily accessible and socially acceptable minimum income fall-back position for everyone. It is shown that a BI scheme, due to its unconditionality, is much more conducive to the ideal of CJ than a conditional scheme of social security. Section 6 elaborates the link between CJ and parasitism. The most serious obstacle to the acceptance of the BI proposal is the parasitism objection: A BI allows some able-bodied citizens to free-ride on the productive

[1] An earlier, shorter, version of this chapter was published under the same title in the *Journal of Social Philosophy*, Spring 2002, 33 (1), 141-161.

efforts of workers. However, absence of parasitism and CJ are mutually conflicting, and to allow parasitism can be seen as a price worth paying to secure CJ. Conclusions can be found at the end.

2. THE ECONOMIST'S VIEW ON COMPENSATORY JUSTICE

Most people think it is fair that work of low status, which is stressful, monotonous, or dangerous is rewarded with higher incomes than jobs which have similar requirements in terms of abilities and training but better working conditions. This idea of CJ is as old as modern economic science itself. Adam Smith, the founding father of this idea, described how '... the whole of the advantages and disadvantages of the different employments of labor and stock must, in the same neighborhood, be either perfectly equal or continually tending toward equality' (Smith 1982, 100). The theory of compensating or equalising wage differentials, as its name indicates, predicts that wage differentials between jobs with similar requirements are such that the sum of monetary and nonmonetary compensation are equalised.

According to Smith (*ibid.*, 101) 'The trade of a butcher is a brutal and an odious business; but it is in most places more profitable than the greater part of common trades.' He adds that '... executioner is the most detestable of all employments'. These may seem prima facie cases of CJ, but it is not as straightforward as at first sight it may seem. It is possible to fill all positions for butchers (or executioners) without any need for compensating wage differentials if their are enough workers who love to be a butcher though the great majority hates it. In a competitive labour market it is perfectly conceivable that the reward for the butcher's trade is even less than the rewards for other, comparable jobs (see Rees 1975, 340). In this sense, compensatory payments can be considered as a kind of wage premium paid for scarce preferences, analogous to wage premiums paid to workers with scarce skills: compensatory payments are just (implicit) market prices determined by labour demand and the preferences of the workers. In general, the need for compensating wage differentials for different jobs in an economy with a high degree of division of labour is much less when preferences are heterogeneous instead of homogeneous.[2]

Since the whole idea behind compensation is to redress utility losses engendered by unpleasant working conditions, it is not unfair that butchers do not receive a compensatory payment. They may well see the remuneration for their labour as a fair day's wage for a fair day's work. At the same time it is efficient to allocate workers with matched preferences to these jobs.[3] It is fair (and necessary in order to

---

[2] Adam Smith implicitly relied on the assumption of homogenous tastes. For instance, Rees (1975, 339-40) writes that: 'There is one important respect in which Smith's theory about the agreeableness of jobs now seems to be incomplete. He writes throughout as though all workers had identical tastes – as if what is disagreeable to one is disagreeable to all... Smith lived in a society which by present American or even present British standards was culturally quite homogeneous. In such a society the assumption of uniform tastes is quite natural... In a more diverse culture, this assumption of naturally uniform tastes may be less relevant.'

[3] See Rosen (1986, 642): 'In contrast to the standard market paradigm, where the identities of traders is immaterial to final outcomes... with whom and for whom one works is generally of considerable importance for achieving efficient labour allocations... the labour market must solve a type of marriage problem of slotting workers into their proper "niche" within and between firms.'

elicit the desired supply of workers for these jobs) to compensate those extra workers needed on the margin for the suffered disutility only to the extent that the number of matches of certain jobs with workers with matched preferences falls short of the number of workers required. However, since equal work is paid equally, workers with 'matched' preferences receive a rent, whereas the workers at the margin do not. The workers on the margin do not have matched preferences, since they are only prepared to accept the job if it pays a compensating wage differential. The rent component of the wage is thus the difference between the actual wage received and the wage which would be necessary to induce a worker, given his next best alternative, to take the job. Although one may find this rent component unfair, it is an inevitable by-product of the principle of equal pay for equal work. If equal pay for equal work is not considered unjust, then these 'compensatory' rent payments are also not unjust.

The size of the rents received by workers with matched preferences is in a quantitative sense not as important as it may seem at first sight. With a reasonable degree of heterogeneity of preferences[4] and a highly diverse job distribution, as is the case in modern economies, the equilibrium compensating wage differential for a particular job is much lower than the average wage premium which would be required if we were to average the compensatory payments needed for *each* worker to take this job. One can think of the average compensatory payment as what would result if we were to pick out persons from the labour force at random and ask them what compensatory payment they need to prefer this job to their next best alternative (see Figure 1). Due to the heterogeneity of preferences and the division of labour, and the matching of workers to jobs, the market will pay much less. The allocational efficiency of the market thus minimises the volume of compensatory payments, and therefore also the volume of rents obtained by the nonmarginal workers. Moreover, while the rents do not count as a social welfare loss, because it is income for workers with matched preferences, a violation of the conditions of CJ will generate *real* social welfare losses, for instance, if we were to force at random workers from the labour force (such as unemployed) to take up the butcher's trade against the equilibrium compensatory payment. Therefore, not only from the point of view of equity (the equal pay for equal work principle), but also from a welfare and efficiency point of view it is very important that CJ prevails.

Figure 1 illustrates the required compensatory payments for one type of job, e.g. the butcher trade, or for one type of job attribute, e.g. to do physical work in the open air. On the abscissa are the compensatory payments for this job and the fraction of the labour force which will choose this job given the compensatory payment offered and given the next best alternative (as we will see later on, workers' next best alternative plays an important role for CJ). The higher the reward, the larger the fraction of the labour force which will prefer this job above their present employment. If only a fraction (P) of the total labour force is required to fill all available positions for butchers, then the compensatory payment OD needed to elicit

---

[4] There cannot be compensatory *rent* payments if preferences are homogenous, since all workers then have equally (un)matched preferences for any type of job.

this fraction (or more precisely, to elicit the labour supply of workers on the margin for this job) is far less than the compensatory payment OC for the 'average worker'. On average they need OC, which is much higher than OD. The allocational efficiency of the market thus minimizes simultaneously the volume of compensatory payments and the volume of rents obtained by non-marginal workers. All workers to the left of P have in a degree matched preferences for the butcher's trade compared with workers to the right of P. Total rents for this group equals the area DEF, and the existence of these rents is entirely due to the principle of equal pay for equal work. The efficiency of the market to minimize compensatory payments for this job can then be expressed as the difference between OC and OD times the number of workers required. This welfare and efficiency gain will be higher (and probably greater than the total value of rents) if preferences are strongly heterogeneous rather than homogeneous.

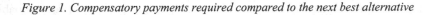

Figure 1. Compensatory payments required compared to the next best alternative

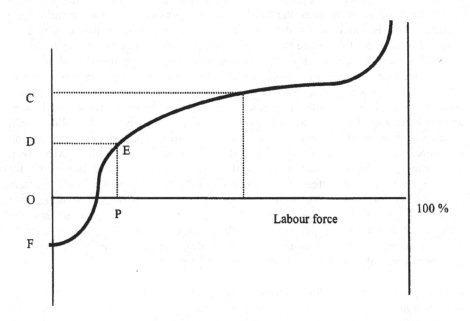

According to the above analysis, the economist's principle of CJ is the principle that each *marginal* worker's pay should exactly compensate for the welfare difference between the job and the next best alternative. The relevant alternative is defined in terms of the set of jobs for which the worker is *readily* employable (e.g., being appropriately skilled) rather than the set of jobs for which the worker is *potentially* employable (e.g., if he had chosen an entirely different educational career).[5] It is

---

[5] In section 5, I argue that the set of options describing the relevant alternative should include a socially acceptable and readily accessible fall-back.

precisely this restriction (readily instead of potentially) that allows the economist's principle of CJ to be compatible with noncompensatory talent-based wage differentials (that is, in so far they do not genuinely compensate for training costs and efforts to acquire these skills). Contrary to the balancing view on CJ (see section 4), which requires low (high) pay for (un)attractive jobs, wage differentials based on morally arbitrary factors as talents and powers are not removed.[6] This is because the principle of CJ underlying the theory of compensating wage differentials does not seek to eliminate all kinds of noncompensatory wage differentials.[7] For instance, talent-based wage differentials and wage premiums for scarce skills may exist along with compensatory wage differentials.

The large number of empirical studies on compensatory payments illustrate that economists have no difficulty in understanding what CJ is about.[8] These studies show that CJ plays an important, though limited, role in explaining wage rates. However, they do not sufficiently emphasise that the occurrence and the level of compensatory wage premiums are strongly dependent on (market) conditions, and on the fall-back position of workers. If the best alternative for a worker doing an arduous job is destitution (having no real access to another and better job), then he may still prefer to take it rather than face destitution. Yet, destitution is an unacceptable alternative, hence not adequate to serve the demands of CJ. Therefore, the levels of compensatory payments will vary with the level of minimum wages, the rate of unemployment, the generosity of and the eligibility conditions on social benefits, etc. This lacuna is taken up in section 5.

3. THE OBJECTIVE APPROACH TO COMPENSATORY JUSTICE

The objective approach tries to determine the need for compensating wage payments by relying on the possibility of making more or less objective judgements, or by reaching a consensus about cases where compensatory payments are required. Baker (1992) proposes to '... set each occupation's rates of pay according to what the typical person (defined by mean or median or mode) would consider to be an adequate level of compensation' (*ibid.*, 109), or, which amounts to more or less the same, '... begin by making a list of the kinds of consideration egalitarians will want to bear in mind in thinking about compensation ("matters of compensation") and then apply this list to particular occupations to establish levels of compensation' (*ibid.*, 110). As with needs, Baker tries to reach the same consensus or background agreement about 'matters of compensation' and their proper levels of compensatory

---

[6] However, the pattern of market wages would reflect the negative correlation between attractiveness of jobs and pay levels if everyone has equal talents. Unequal talents and scarcity-based wage premiums thwart this outcome.

[7] In this respect the principle of CJ underlying the economist's view is far less comprehensive than those of the other two approaches.

[8] See Hwang *et al.* (1992), Killingsworth (1986), Smith (1979). Hersog and Schlottmann (1990, 469) estimate the implicit price for a work-related fatal injury between $1.7 to 2.3 million.

payments.[9] According to Baker's proposal, a butcher would be entitled to the average required compensatory payment, even if all workers who actually were butchers liked their jobs.

However, the analogy between special needs which require extra resources and special objective or intersubjective assessed burdens which require compensation breaks down. Extra resources for needs are given to the handicapped in order that they can attain the same standard of living as other persons, while compensation is 'an inherently "welfarist" concept... the idea is one of supplying something (usually, though not necessarily, money) which is supposed to provide a source of utility equivalent to that lost' (Barry 1992, 135). The difference between the entitlements of needs and burdens is that the first refers to the objective costs to regain as far as possible normal functioning (not to compensate for the disutility suffered from the handicap), while the latter refers to a monetary compensation equivalent to the utility loss experienced by workers for what they see as unpleasant job attributes. The extra resources for special needs depends on objective costs, while the amount of compensation depends on the preferences of workers. Baker wishes to compensate according to the average preference, or some objective standard of burdens determined by a majority communal judgement, but in doing so he has to face two problems. Firstly, with a high degree of division of labour, the number of 'matters of compensations' would be endless. Secondly, even if this number were limited, there would be an enormous difference between the established objective compensations (if consensus can be reached at all) and the subjective cost of job burdens because preferences are heterogeneous. The objective approach is therefore impracticable[10] (it requires an insurmountable amount of information on working conditions in different jobs, to establish the basis upon which consensus must be reached) and it would be inefficient (because it does forego the welfare and efficiency gains of leaving the determination of compensatory payments to the market). Baker's proposal would make more sense if both jobs and preferences were homogeneous and in the absence of a labour market (for instance in a primitive economy). However, with a high degree of division of labour and a strong heterogeneity of preferences, a well-functioning labour market will reveal in an efficient way where compensatory payments are required. This is only the case if the conditions required for CJ are met, so that the market forces responsible for achieving a state of CJ can work unhampered.

4. THE BALANCING APPROACH TO COMPENSATORY JUSTICE

The balancing view on CJ primarily objects to the phenomenon that the most attractive, secure and pleasant jobs are usually well paid, and most tedious and insecure work is as a rule badly paid. Instead, those of this view would like to see that the wage distribution is the other way around, namely that wage differentials

---

[9] '... this degree of background agreement does not represent identical preferences but only a shared belief that a certain set of differentials constitute a set of reasonable levels of relative compensation with respect to a reasonable list of objective benefits and burdens.' (*ibid.*, 111).

[10] Admittedly, if an approach proves impractical this does not show that it is (conceptually) wrong.

reflect nonmonetary disadvantages between jobs. Typical for this approach is that it takes the ideal of CJ as a balancing principle of just income distribution, not allowing that more capable workers in well-paid jobs have better working conditions than less capable workers in low-paid jobs. In the economist's view on CJ, '... workers with greater earning capacity would "spend" some of it on more on-the-job consumption. This is the fundamental reason why low paying jobs tend to be the "worst" jobs' (Rosen 1986, 671).[11] Contrary to what the balancing approach requires, even if all conditions of CJ according to the economist's view are met, it will be observed that CJ does not lead to a kind of balanced equilibrium of pleasant jobs with low pay and unpleasant jobs with high pay. If working conditions are normal (luxurious) goods, which can be priced, bought and sold, it implies that when earnings go up, the demand for better working conditions will (more than) proportionally rise. Part of the total pay package takes the form of holidays, fringe benefits, emoluments, pension plans, job security, promotion opportunities and other indirect rewards.

My criticism of the balancing approach is twofold. Firstly, seen from the economist's view on CJ, it carries the message too far: only under very restrictive assumptions will wage differentials fully reflect job (dis)advantages and (dis)amenities. This pattern only emerges in the imaginary case of homogeneous preferences, identical talents and skills and a perfect market so that differential wages are exactly the mirror image of the non-pecuniary (dis)advantages (see below). As soon as the assumption of identical skills and talents is removed, scarcity payments will determine relative wages along with compensatory payments (which in turn can be seen as a payment for scarce preferences).

Secondly, there is no institutional design available under which this kind of CJ can be achieved (to be sure, the unfeasibility of the balancing view is not an objection to its conception of CJ). In this section I will discuss two authors (Dick (1975) and Carens (1985)) who say that CJ requires a negative correlation between wages and working conditions, instead of the positive correlation we see in the real world. In contrast to the objective view, both authors acknowledge the role of the market in achieving CJ.

In order to solve the problem why the majority of higher-paid jobs are more pleasant than the lower-paid ones, Dick (*ibid.*, 266) splits the market wage into two components, the 'transfer earnings' and the 'economic rent'. Transfer earnings are equal to '... the amount necessary to keep him from transferring to another job', while the rent component in the market wage is equal to '... an amount in excess of what is necessary to induce a worker to take a job. The element of economic rent is thus that portion of a worker's pay which from a psychological standpoint does not act as an incentive to perform but is instead supererogatory' (*ibid.*, 268). The difference between the actual wage and the transfer earnings represents for Dick a kind of undeserved rent. If, in one way or another, the rents can be withheld at source, so that wages only reflect transfer earnings, then this '... provides the most

---

[11] This statement makes clear that the economist's view on CJ, here represented by Rosen, does not require that earnings exactly balance nonmonetary disadvantages.

powerful and important ground for justifying differences in income.... the necessity of offering monetary compensation to bring forth a required supply of labor does not imply that such payments are also just. But there is good reason to believe they are' (*ibid.*, 264) and compensatory payments '... offer no special problem from a normative point of view. They are deserved because they are compensatory...' (*ibid.*, 266).[12]

The problem here is that although this distinction can be made on a conceptual level, it cannot be used on a practical level in order to establish a more just income distribution by withholding the rent component. As argued above, the rent component received by workers with matched preferences (the intra-marginal workers) is the inevitable by-product of the principle of equal pay for equal work. Whether transfer earnings are determined on an individual level or per job, in both cases the competitive market wage must be equal to the transfer earnings of the workers on the margin (those with the least matched preferences but still inclined to take the job at a particular level of transfer earnings) to elicit the desired supply of these workers. There is no way to withhold at source, or to tax away, the rents received by intra-marginal workers: on an individual basis informational constraints and the principle of equal pay for equal work block the withholding of rents at source, while on a job-like basis there are no rents to withhold.[13]

To see the impossibility of using the distinction beyond the conceptual level, suppose the procedure is applied either on an *individual basis* (with all its informational problems) or on a supra-individual *job-like basis*. In the individual approach,[14] transfer earnings for the same job will differ among individuals because preferences are heterogeneous. Since competitive forces will equalize pay rates for equal work, the economic rent component in the wage will differ among workers doing the same work. Clearly, it is not the case that 'it should be possible to tax away or withhold at the source the element of economic rent in everyone's pay and still elicit a reasonable amount of the desired sort of work' (*ibid.*, 270). How do employers or tax authorities know to what extent workers have matched preferences, and subsequently manage to withhold or tax them on a selective basis?

For the desired supply of workers for a job to come forward, Dick is obliged to follow the procedure of splitting the market wage into transfer earnings and economic rent on a job-like (or job for job) basis. In that case, one may wonder where in a well-functioning labour market the difference between the market wage and transfer earnings leading to an economic rent may come from, given the constraint that the desired supply of workers for a job must come forward. Employers offering a market wage equal to the transfer earnings on a job-like basis

---

[12] This is, of course, only true in the unlikely situation that all workers with the same job have the same (homogeneous) preferences towards this job, and therefore find it equally undesirable, harmful or degrading.

[13] The argument here only pertains in a perfect labour market since it may well be the case that some occupations' rate of pay is far above the required level of transfer earnings (that is, what can be justified on compensatory grounds), e.g., occupations in which there is a cartel, or managers who can set their own levels of pay. To eliminate these rents would require different institutions, a problem that will not be discussed.

[14] Which is, I think, Dick's approach since he formulates his proposal and the distinction between transfer earnings and rents in terms of individual workers.

are able to attract the desired supply. The job-specific economic rent is nil, although there is a rent component for some workers with matched preferences.[15] Even if a way could be devised to extract rents, it would violate the principle of equal pay for equal work. Although Dick's proposal may make sense on the conceptual level, it lacks a clear-cut institutional framework which can effectuate this type of CJ.

Carens (1985) connects the conceptual issue what CJ is with the institutional issue how to achieve it: '... the theoretical ideal of compensatory justice cannot even be adequately articulated without reference to a particular model of how that ideal should be institutionalized' (*ibid.*, 46). After this statement one would expect a discussion of the conditions or institutions required for implementing CJ,[16] but Carens turns his attention to the first issue: how CJ would fare under ideal circumstances in which all would work against individual-specific transfer earnings (and consequently each would give voluntarily to the treasury the rent received), what ought to be compensated (sacrifices, burdens, efforts, risks, training costs), which metric to be used and discussion about finding a method to determine how much these compensations need to be.

Here Carens addresses the problem of the positive correlation between wages and working conditions: 'In the real world... people vary in natural talents and in the skills they are able to acquire, and competition is far from perfect. Indeed, in existing markets it is probably fair to say that the worst jobs tend to be the worst paid and the best paid jobs offer far more than could ever be justified on compensatory justice.' (*ibid.*, 56). Carens acknowledges that a large part of the actual income distribution is not determined by only CJ, that is, a large part of the income inequalities are noncompensatory[17] and that the attempt to reverse the positive correlation into a negative one is bound to come into sharp conflict with the demand for efficiency. If wages must reflect compensations, they will reflect burdens, efforts, overtime work and so on, but not rewards for scarce talents, skills or differing (marginal) productivities of workers. Untalented, hard workers would receive a higher wage than talented workers who take it easy, despite the higher productivity of the latter. This tradeoff between balancing CJ and efficiency is rightly stressed by Carens and his rough conjecture is that '... the most satisfactory way of institutionalizing the ideal of compensatory justice would be to try to

---

[15] Suppose two groups of workers which differ in the type of skill but not in the level of skill (hence, equal training costs) and that only one type is scarce in supply. A worker with the scarce skill will receive a higher wage, but this cannot be classified as a rent in Dick's terms because this scarcity premium is 'necessary to keep him from transferring to another job' requiring similar skill.

[16] Although Carens acknowledges that a well-functioning market is one of the most important institutions for achieving CJ: 'Compensatory justice has a powerful intuitive appeal which can be captured and expressed by pointing to our strong sense that, whatever distribution of income would emerge in a perfect market among individuals with equal abilities, that distribution would be fair' (*ibid.*, 67). As outlined in section 2, the important role of the market in achieving CJ is that it provides a mechanism for bringing about implicit prices for (un)desirable job attributes, for allocating workers to jobs (to solve the marriage problem of allocating heterogeneous workers to the right jobs) and therefore for minimising the amount of compensatory payments required.

[17] The paradigm example given by Carens is the income of superstars. In Borghans and Groot (1998) it is shown that these superstar wages are indeed largely composed of rents due to monopolistic power of number one positions in media-activities.

equalize incomes as much as possible (within the constraints of efficiency)' (*ibid.*, 40). The problem of the balancing approach therefore is that the implementation of this ideal would cause prohibitively high efficiency losses. These efficiency losses reduce the resources available to improve the (fall-back) position of the least productive workers who do not have unattractive jobs.

For this reason, I propose to achieve CJ by providing a decent fall-back minimum income unconditionally to all. This being only one determinant of the income distribution among others (factors like different talents, (marginal) productivities and scarcities, adequate provision of special needs, etc.), it can be supplemented with other criteria of a just income distribution (e.g. maximin equality, see below). The price to be paid for this more modest strategy of CJ is that the view that all income inequalities must be compensatory in nature have to be dropped, and that the problem of the positive correlation remains.[18]

In the following section I will defend the view that the most important element in the institutional framework designed to attain CJ is to provide a decent minimum income guarantee unconditionally (that is to say, a socially acceptable no-work option) to all citizens. Note that to some extent a generous level of BI (or NIT) nicely fits the maximin rule of income distribution.[19] With both maximin and BI, the noncompensatory payments (reflected in the – remaining – positive correlation between jobs and rewards) would be considered justified as long as it has the effect of raising the level of BI provided to all.

5. THE CONDITIONS OF COMPENSATORY JUSTICE: THE ROLE OF THE SOCIAL SECURITY SYSTEM

So far we have scarcely paid attention to the influence of the institutional framework of the economy on CJ. The importance of a proper functioning labour market was only mentioned, and nothing very specific was said about the impact of alternative systems of social security on CJ. This section discusses the importance of the type of social security on CJ. As a beginning, consider the following rather lengthy statement of Mill:

> ... it is altogether a false view of the state of facts, to present this [the view of Adam Smith cited in section 2 above, LG] as the relation which generally exists between agreeable and disagreeable employments. The really exhausting and the really repulsive labours, instead of being better paid than others, are almost invariably paid the worst of all, because performed by those who have no choice. It would be otherwise in a favourable state of the general labour market. If the labourers in the aggregate, instead of exceeding, fell short of the amount of employment, work which was generally disliked would not be undertaken, except for more than ordinary wages. But when the

---

[18] The fact that '... most income inequalities are not compensatory but additive in the sense that they add to the advantages in working conditions, intrinsic interest, and so on...' (*ibid.*, 65) is a problem only if all income inequalities must be compensatory in nature. Workers with greater earning capacity will always 'buy' more in terms of good working conditions than workers with a low productivity level, for the same reason that they are living in more comfortable houses.

[19] For the link between maximin and BI, see Van der Veen (1991, chapter 2) and Van Parijs (1995, sections 4.1 and 4.2).

supply of labour so far exceeds the demand that to find employment at all is an uncertainty, and to be offered it on any terms a favour, the case is totally reverse. Desirable labourers, those whom every one is anxious to have, can still exercise a choice. The undesirable must take what they can get. The more revolting the occupation, the more certain it is to receive the minimum of remuneration, because it devolves upon the most helpless and degraded, on those who from squalid poverty, or from want of skill and education, are rejected from all other employments. Partly form this cause, and partly from the natural and artificial monopolies,.... the inequalities of wages are generally in an opposite direction to the equitable principle of compensation erroneously represented by Adam Smith as the general law of the remuneration of labour. The hardship and earnings, instead of being directly proportional, as in any just arrangement of society they would be, are generally in inverse ratio to another (Mill 1968, 383).

Obviously, the conditions for CJ were not met when Mill wrote this passage. No compensatory payments were made because some workers (the most helpless, degraded, destitute, unskilled, otherwise unemployed) had no real choice (they had to take what they can get). Is Mill's observation still valid today? I think most people have some idea of the hardships endured by most workers in Mill's day and of the harsh kind of Poor Law regime in the nineteenth century. The social benefits today, both in absolute and in relative terms, for unemployed, sick, disabled, elderly and poor citizens (and children) are much more generous than they were in the Victorian era. The improvements made with respect to CJ are indeed related to the rise of the welfare state, that is, with the institutions under which the market economy now operates. In what follows I will describe a number of factors which are important for CJ and evaluate the effects of the present conditional system of social security compared to the imaginary case of a BI system on CJ.

Free educational and occupational choice, and free entry to and free exit from jobs, are institutional conditions for CJ which immediately spring to mind. Normally free educational and occupational choice is subsumed under the heading, and discussed in the context of, freedoms or basic liberties (see also the appendix). The argument here is not that free career choices are an essential component of our set of liberties, they definitely are, but that they are required from the point of view of CJ. Since we do not fully know workers' preferences, nor their heterogeneity, or the (dis)advantages that job attributes offer to workers, we must leave the allocational problem of slotting workers to educations and jobs to the free traffic on the labour market. Some trainings or jobs may temporarily offer more than average returns compared to similar trainings or jobs. This will be a short-term phenomenon if the relevant conditions of CJ are met: 'If in the same neighbourhood, there was any employment evidently either more or less advantageous than the rest, so many people would crowd into it in the one case, and so many would desert it in the other, that its advantages would soon return to the level of other employments' (Smith 1982, 100).

Ever changing conditions on the labour market, with some industries declining and others expanding, put a lot of pressure on the educational system to provide the required supply of skilled workers. Unless workers with different educational backgrounds can easily be substituted one for another, it presupposes a kind of perfect foresight of students to make the right choices. A BI facilitates the retraining of unemployed or dissatisfied workers. If, by chance, one occupational group is short in supply, it will be easier for workers (or the unemployed) under a BI scheme to redirect their labour market career in this direction by schooling. They can quit their present jobs, and live from their savings and the BI during the time needed to qualify for some other job. The BI may serve as a long-term device for facilitating this condition of CJ, whereas under conditional social security one is usually not entitled to social benefits while engaged in formal schooling or training.

The condition of full employment is more controversial and complicated. One may expect better chances for CJ in a state of full employment than in a state of involuntary unemployment. I think this is the main reason why one is inclined to think that the income distribution emerging in a perfect market with equally talented workers but differing tastes is fair is because markets clear under perfect competition, and no unemployment results. Each worker can then truly balance the package of the wage and non-monetary (dis)advantages attached to one job with that attached to other jobs and differential wages reflect compensations for unattractive job attrbutes. Employers are forced to pay positive wage differentials for job attributes for which workers with matched preferences are in scarce supply.

The perspectives for CJ becomes worse with longlasting and large scale unemployment. Most welfare states have to face the problem of high unemployment and this will probably still be with us in the near future. The unemployed may prefer to do disagreeable work at low pay rather than remain unemployed and see their skills atrophy. The extent to which the demands of CJ are violated under these circumstances depends strongly on the generosity of the social security system. It depends not only on the level and duration of unemployment and social assistance benefits, but also on the conditions for receiving them.[20]

5.1. Compensatory justice and conditional social security

Crucial for the degree of CJ that is achieved is that all the present social benefits are not unconditional. One must really be sick or disabled, young or old, poor or involuntarily unemployed to be entitled to one of these social benefits. The duty to work is one of the cornerstones of the present social security system. A system of work- and learnfare of the type towards present social security is moving, is largely based on this duty. A really conditional scheme cannot do without the principle that able-bodied adult citizens have a duty to work. Against the right to a social benefit, there is a duty to accept labour. Graduates and dissatisfied workers cannot ask for social benefits simply because they prefer being unemployed for some time rather than employed. At most, one can qualify for a social assistance benefit for which one has to pass a means-test. However, a great majority of the employed fail to pass the means-test due to what they have saved in the past. Those workers who can pass the means-test do not receive the full amount of social assistance, because when they

---

[20] According to Rees (1975, 339) 'The higher the level of liquid assets of the unemployed and the stronger the system of unemployment insurance or other forms of income maintenance, the weaker the tendency for the unemployed to take disagreeable jobs at low wages. However, it should still be true that compensating differentials will be best observed in periods of full employment.'

have left their job voluntarily, benefits are cut for a limited period. Thus the income of the fall-back position for workers who do not have immediate access to another job, and with an amount of savings higher than permitted to become entitled to a social assistance benefit, is zero. For those who could pass the means-test the income of their fall-back position is positive, but less than the level of social assistance benefit, which is already at the social minimum. Whether one can pass the means-test or not, making oneself unemployed because of dissatisfaction with the total pay package of the job, is not a very attractive option.

The other alternative is to look for a job elsewhere. However, if some employers on nonclearing segments of the labour market are not forced to pay compensatory payments for particular job attributes, other employers in the same or neighbouring segments are in the same position and are not forced to do so either. This may explain why all supermarkets pay more or less the same wage rates and offer more or less the same conditions of employment to their cashiers and shelf-stockers. So, although competition among workers for jobs and among employers for workers will result in CJ on those segments with no excess supply of qualified workers (comparable to the degree of CJ that would result in a state of full employment), it does not lead to CJ in those segments for which there is an excess supply of workers. In summary, we may expect CJ only in those segments where the labour market clears. Where it does not, it depends on whether workers have ready access to an acceptable alternative.

The position of the unemployed differs depending on whether they receive a non-means-tested unemployment benefit or a means-tested social assistance benefit. The level of benefit functions as an important mechanism for forcing employers to pay compensatory payments if workers were entitled to benefits for as long as they wanted. However, the duties, threats and stigmas imposed on social security recipients actually precludes this kind of CJ. The unemployed face a serious penalty for refusing jobs offered by job centres, since part of their social benefit will be curtailed. The work-test, if really applied, reduces the power of the unemployed to ask compensatory returns for unattractive labour. Employers can use the threat of potential cuts in social benefits to keep wages low. Moreover, in as far as the unemployed feel stigmatised and humiliated because of receiving social benefits, the utility attached to the unemployment status is lower than it would have been if stigmatisation and humiliation were absent (as is the case under a BI regime) and have therefore less credible power to ask for compensatory payments, since their fall-back position is worse (see also the appendix).

An austere workfare scheme, imposing really severe conditions on social security eligibility and penalties on non-participation in schooling programmes and (unpaid) social services, fares even worse in this respect. It worsens not only the position of the unemployed, but also of all other low wage workers. Employers of low wage workers know that the alternative for a worker is an austere workfare regime and this worsens workers' terms of negotiation. Increased labour supply of unemployed, combined with worsening terms of negotiation of low wage workers, exerts a downward pressure on wages. This may result in an increase in wage inequality, the opposite of what is required for CJ.

As previously argued, conditional social benefits have their shortcomings for achieving CJ. Minimum wage and employment conditions legislation may provide a remedy. Unfortunately, seen from the perspective of CJ, both strategies have their own drawbacks. Firstly, minimum wage legislation under conditional social security can be taken as a signal that the conditions of CJ are not fully met. The need for minimum wages suggests that the fall-back position of low wage workers is not strong enough to let wages on this segment of the labour market be determined by demand and supply forces. Minimum wage legislation ensures that each worker can earn a living wage.

Secondly, minimum wages are non-discriminatory between low paid bad jobs and low paid pleasant jobs. Some jobs may attract enough workers even if the wage is below the minimum wage. Since the minimum wage legislation acts as a binding constraint in some segments of the labour market, labour demand for these jobs is lower than what it would have been without the constraint and workers with these jobs may receive a higher wage than required on compensatory grounds. This indiscriminatory nature of minimum wages between good and bad jobs is in sharp contrast with the wide-spread dispersion of both workers' preferences for jobs and the wide-spread dispersion of job attributes. Whether jobs and job attributes are attractive or not is in the last resort a subjective judgement. Under a BI there is no need for minimum wage legislation, since workers are protected already by means of the free access to a BI.

The second strategy, legislation on conditions of employment, can only play a limited role. Workers may lack complete information about the dangers of working in a polluted environment or working with materials harmful to health. It is very important to protect ignorant workers against these hazards. However, the need for *comprehensive* legislation becomes much less pressing under a BI regime. Much more can be left to the forces operative on the labour market. There is no need to have legislation on the (social security) rights of flex workers, the number of vacation days, working hours, compulsory retirement, etc. All these issues can be left to the market, since the BI provides each potential worker with substantial bargaining power against employers. Slotting workers with a high preference for labour into jobs with a long working week and few free days in return for higher pay is not only efficient, but is also compatible with the demands of CJ if this preference is scarce in supply. In short, if a decent fall-back income is provided to everyone, one has the option to make the labour market really flexible.

Finally, protecting workers under a conditional scheme, by either making the social benefits more generous or by raising the minimum wage, may have important adverse effects on the level of employment. Raising the means-tested benefit enlarges the poverty trap, while an increase of the minimum wage makes it more difficult for employers to offer profitable employment for low productivity workers. Both measures have a positive effect on the rate of unemployment, and this has, ceteris paribus, a detrimental effect on CJ.

5.2. Compensatory justice and basic income

It appears that, due to its unconditional nature, this *prima facie* case of the BI for CJ depends to a large extent on the level of BI. The primary role of a BI for the attainment of CJ is to ensure that unemployment is an acceptable alternative without distorting competitive forces on the labour market for low wage labour. I will assume that this level is not higher, and perhaps somewhat lower, than the present level of the social assistance benefit.

By providing a BI unconditionally, the income and utility[21] which potential workers derive from this no-work option serves as a floor. Provided the level of the unconditional grant is around subsistence level, all labour supplied is supplied voluntarily. The no-choice position, which was of paramount importance in Mill's statement, and is of paramount importance for the absence of CJ, is eliminated. With a substantial BI around the level of the social minimum, or even somewhat below this level, nobody is forced to accept a job because of severe poverty. The phenomenon of excess supply for, say, low-skilled, unpleasant jobs can no longer be a real obstacle for CJ, since the low skilled workers have the option to refuse until it is better paid to compensate for its unpleasantness, or until its conditions of employment are improved. The damaging effect of the reserve army of unemployed on CJ implicit in Mill's statement is blocked.[22] Indeed, one of the likely effects of a switch towards a substantial BI is either a rise in the net income of workers with unappealing, routine work, or a significant improvement of the employment conditions in these segments of the labour market, or both, depending on the costs to be incurred in each alternative by employers in order to recruit enough workers.

In Mill's statement there is a clear reference to the limited choice open to those who are least productive. With a BI, the range of choices open to those with the lowest earning capacities becomes wider:

> ... the less valuable one's talents, the more one will gain, in terms of opportunities for fulfilling occupations, from an increased BI. If your earning power is low, there may be only very few occupations that will give you a subsistence income with no BI or a low one. With a substantial BI, your choice will be far greater because you need less income, if any, to reach subsistence (Van Parijs 1995, 124-25).

The BI can be used by low skilled workers as a wage subsidy to price themselves into (more interesting or satisfying) jobs.

Still, wage inequalities under a BI scheme will partly, perhaps even largely, be determined by differences in talents and by the relative scarcity of some skills compared to others. Despite the higher relative wages to be paid by employers to elicit enough labour supply for unpleasant jobs, noncompensatory payments will

---

[21] The lower is the utility level of the no-work option, the less employers have to pay potential workers to bring them up to that level (technically, their reservation wage will be lower).

[22] A remarkable difference between a conditional and unconditional scheme of social security is the meaning of full employment. Under the first it must be interpreted as full voluntary and involuntary employment, while under the latter it can only be full voluntary employment. In the conditional scheme the government cannot (or should not) tolerate a situation in which some are voluntarily unemployed while receiving social benefits. However, in the unconditional scheme it is perfectly imaginable that some people may prefer to remain unemployed while living from the BI only.

persist. Even if all conditions of CJ are met, the positive correlation between earnings and non-monetary advantages will remain, although certainly less than observed by Mill, and probably also than we observe today. However, it is likely that this positive correlation is less strong the higher is the level of BI, since the high income earners are net contributors to the BI scheme, while low income earners are net recipients.

6. COMPENSATORY JUSTICE AND PARASITISM

Compensatory payments for any job are strongly dependent on potential workers' best alternatives. For those without a job under a BI scheme, the fall-back position is largely determined by the level of the BI. However, to abstain altogether from work and to live from the BI is seen by many people as parasitic. Even more important is that labour income, in contrast with the BI, can be interpreted as a compensation for giving up leisure time.[23] If we want to advocate a BI for reasons of CJ, should we not begin by pondering what the compensatory nature is of the BI itself? If someone receives a BI and does nothing in return, can we still maintain that the BI is compensating something?[24] I will try to defend the BI against this objection in relation to CJ by means of two independent arguments. Firstly, CJ and absence of parasitism are mutually conflicting ideals in times of unemployment. Secondly, using a neutral work ethic, expressed in slogans like 'those who do not work, shall not eat', 'you can't expect something for nothing' and 'there's no such thing as a free lunch', it can be argued that, *from the viewpoint of CJ*, only the *in*voluntarily employed have reason to complain about the voluntarily unemployed who choose to live from others' labour. It can easily be shown that the position of these involuntarily employed will improve by switching from a conditional to an unconditional regime.

With full employment we can have both absence of parasitism and CJ. To anyone who applies for a social benefit, we can say, here you have a job instead. Unfortunately, one cannot have both absence of parasitism and CJ at the same time in a state of unemployment. These are mutually conflicting ideals. Absence of parasitism (no improper use of social benefits) can only be reached by means of a duty to work and the corresponding potential threats of cuts in social benefits as in a conditional scheme of social security. However, the stronger these duties and threats are applied to the (willing and unwilling) unemployed, the weaker their terms of negotiation against potential employers and the less scope workers have to command compensatory payments. The absence of a threat of a cut in social benefits, and the absence of a duty to accept work under the BI scheme, strengthens workers' fall back position, but the price to be paid is that of parasitism. It is not that

---

[23] On the same footing, the unemployment benefits can be interpreted as the compensation for being involuntary unemployed because no jobs are available and for which one has paid insurance premiums in the past.

[24] An elaborate and positive answer is provided by Van Parijs's (1995) justification of a BI. A BI partly reflects the per capita value of scarce external means of production, whether natural or produced, plus the per capita value of scarce job assets. In a just society, as defined by Van Parijs, every citizen is at least entitled to a BI equal to the sum of these per capita values.

we have to make do with the parasitism argument, but we have to realise that the struggle for a state of no-parasitism in times of unemployment, in order to reduce improper use of social benefits, has its social cost in terms of less CJ. We saw in section 2 that the social welfare loss of allocating workers to jobs at random, instead of leaving it to the market's allocation process,[25] is considerable. Unemployed people who are forced to take any job offered, as it is under a strict applied duty to work, are more or less at random allocated to jobs.

The above argument points to a tradeoff between CJ and absence of parasitism. The second argument tries to weaken further the parasitism argument. It first restricts the number of workers who have reason to complain about parasites on compensatory grounds. Although all workers sacrifice leisure when working, only for a part (the involuntarily employed) this sacrifice is not fully compensated. Next it shows that these workers will gain from a switch of a conditional welfare state to a BI scheme.

The tenability of the assertion that to work is a duty and that parasitism is wrong rests on the strength of the belief that labour is a sacrifice. If all of us would like to work, irrespective of the income, a duty to work would be pointless. In this case the problem would not be to push willing workers, but to provide enough jobs. The rationale behind the duty to work from an CJ point of view is that work is a sacrifice, and labour income the compensation. However, many workers do not see their job as a complete sacrifice. Labour is not always *labeur*. Many enjoy their work, even if only because of its intrinsic value. Even when they do not always enjoy their work, the wage they receive does more than compensate the sacrifice the work entails. Many unemployed do not see work as a sacrifice either, but as a privilege due to the intrinsic worth of work and to the higher income. Workers for whom work is not a sacrifice, or for whom the sacrifice is more than compensated by the wage, have no moral right to demand sacrifices from those unemployed through no fault of their own, who also want a job because of the privileges it provides. Still, those for whom work is on balance a real sacrifice, the *in*voluntary employed (e.g. the cashiers and shelf-stockers) rather than the voluntary employed, have reasons to complain against the voluntary unemployed who choose to live from other's labour. The involuntary employed are morally in a privileged position to demand that the neutral work ethic is respected.[26]

---

[25] That is, slotting workers with more or less matched preferences to these same jobs by means of price signals.

[26] High paid workers with interesting jobs may nevertheless object: from the perspective that just because they like their job and earn a lot while others do not like their jobs, earn much less or are unemployed, it does not follow that it is legitimate to tax part of the income of the former and to transfer it to the latter group. This conclusion would only follow if one requires that sacrifices are exactly mirrored by monetary rewards (this balancing approach is rejected). How much each one earns, and how much one likes to work, is irrelevant if there is full employment against market clearing wages. However, this view is rather problematic if labour markets do not clear. Then there will always be some involuntary unemployment (e.g., due to efficiency wages, or to collective wage bargaining) while those who work receive higher than market-clearing wages.

It can easily be shown that a BI will particularly improve the position of this group of involuntary employed.[27] The involuntary employed are those who hold jobs which either have a low intrinsic worth or for which the wage offered does not compensate the associated sacrifice. One of the consequences of the implementation of a substantial BI is that the relative wages of unattractive work compared to attractive work must rise in order to induce recipients of a BI to accept these jobs. They will also have more bargaining power to demand better conditions of employment, such as job rotation. Finally, low wage workers will be net recipients under a BI regime (whereas they are net contributors under conditional social security, because they pay taxes and do not receive a benefit), while those with a high earning capacity and for whom work is attractive will be net contributors. If those with the lowest paying jobs also have the least attractive jobs, then this redistributive effect of a BI is fair in so far the rationale underlying the neutral work ethic rests on the premise that working entails a sacrifice.

To conclude I want to give two further, more speculative, considerations related to parasitism and CJ. With an equal social minimum, the balancing budget tax rate of the BI scheme will be higher than the average tax rate of a conditional scheme with categorical benefits. The question arises of whether it is fair to tax workers with a high preference for consumption over leisure in order to finance a BI scheme. Elster (1989, 215) sees this as unjust: 'People who chose to work for an income rather than to live in a commune on the universal grant would have to pay higher taxes in order to support those who took the other option. They would think, correctly in my opinion, that they were being exploited by the other group.' To this objection that a BI amounts to exploitation[28] of those with a strong work ethic by those who are work-shy, the advocates of a BI may answer the following: One's work ethic is a private affair. It is one preference among others, like one's preference for high consumption. These preferences do not have any a priori claim that the institutional framework of society must be such that their satisfaction is made more easy than competing preferences.[29] Only the *average* work ethic is a relevant variable for any kind of social security system, because social security must be financed by means of the contributions made by those who perform paid work. If a (substantial) BI turns out to be economically unfeasible, then this would indeed tip the balance in favour of conditional social security. But suppose a BI scheme is feasible – at a level where enough persons want to convert leisure time into additional money income on top of their BI by means of paid labour – then it might

---

[27] Winners can also be found among the now involuntary unemployed if they can find paid work more easily under a BI regime (e.g., due to the absence of a minimum wage and the possibility to use the BI to price themselves into a job), among low income earners (due to the redistributive effect of a BI), and last but not least, among those with a high preference for leisure (e.g., the parasites and non-needy bohemians).

[28] Note that this exploitation objection is subtly different from the parasitism objection mentioned above. In the latter, the reciprocity-based duty to work (those who can work should not be entitled to social benefits) is at stake, while in the former the lower standard of living of workers in virtue of supporting the parasites is involved.

[29] This may seem strange given the important role of preferences in CJ according to the economist's view. However, in this view no particular preference has an a priori claim on compensatory payments, since it depends on overall demand and supply conditions which preferences turn out to be scarce.

be justified to allow some parasitism at the expense of tax payers to serve the interests of the low wage workers. To insist on a conditional scheme to do justice to the disgruntled workers who feel a duty to work, not wanting to live off others labour, foregoes the advantages of a BI scheme in terms of CJ.

The optimal taxation literature shows that with a universal grant or BI, some part of the labour force will not perform paid labour due to low earning capacities. This is simply because if marketable skills are of little value on the labour market, the sacrifice (the effort and time one has to expend) when taking up paid labour is not compensated by the intrinsic and extrinsic value of paid labour. Admittedly, it is a big step to go from the intuitive view that every able-bodied, non-aged citizen has to incur some sacrifice to earn a living, e.g. by giving up leisure time, to the view that those for whom the sacrifice is too large in comparison with what they can receive in return are not obliged to do so. Nevertheless, it is widely known that in a conditional scheme with a high withdrawal rate and a duty among recipients of social security to accept job offers, there are a large number of recipients who are unwilling to cooperate (that is, to take a job as soon as possible). Seen from the point of view of CJ, this unwillingness may partly be explained by the fact that in such a conditional scheme the requirements for CJ are not fully met, which implies that for certain groups the sacrifice of taking up paid labour is not duly compensated.

SUMMARY AND CONCLUSIONS

In the economist's view on CJ, only the workers on the margin are exactly compensated in welfare terms for their loss of utility compared to their next best alternative, whereas all non-marginal workers (perhaps liking their jobs) are being overcompensated. This outcome is however more efficient than what would follow from an objective or balancing approach to CJ. The chief criterion to be used to assess whether the conditions of CJ are met or not, is whether people have the real choice or no other choice when they take a job. Having a real choice requires an acceptable alternative. If workers lack the real freedom of choice, the prospect of CJ is threatened. The best guarantee for CJ is a decent fall-back position readily accessible to all. According to this view, with the emphasis on the required conditions, it is easier to make comparisons of CJ between different schemes of social security than to measure the degree of CJ.

It was argued that although the objective and balancing approach may have coherent conceptions of CJ, they are infeasible and impracticable. The economist's view on CJ is the most practical conception of CJ: CJ is achieved when each job's rate of pay exactly compensates the worker on the margin for the disutility suffered compared to her next best alternative. However, this criterion is incomplete since it leaves out what the alternative is, and disregards the other conditions which are required to achieve CJ. There can still be a blatant violation of CJ if the alternative or fall-back position (and hence workers' bargaining power) is weak. It suffices to say that the alternative must be socially acceptable. Full employment is usually a sufficient guarantee, and involuntary unemployment the primary threat, to CJ. The duty to work, the need to resume work as soon as possible, the threat of a cut in benefits and the stigmas attached to benefits under a conditional scheme, are factors that are very likely to impair the conditions of CJ or hamper the market forces which bring about CJ. The more the present welfare state is moving away from generous welfare provisions towards an austere workfare scheme, the more serious this danger becomes. This tendency might partly explain the rising income inequality in the US during the last decades. The advantage of a BI scheme is that by providing unconditional access to subsistence income to all, without means- or work-test, it precludes coercive pressure towards the (un)employed to take jobs at rates of pay below what would be required on grounds of CJ. This has far-reaching consequences for the degree of CJ that can be attained in the end: although the positive correlation between good jobs and pay levels is not eliminated, it will probably become much lower.

APPENDIX

RAWLSIAN JUSTICE AND FAIR EQUALITY OF OPPORTUNITY

The purpose of this appendix on Rawlsian justice is twofold. Firstly, a brief outline is given of Rawls's principles of justice and the accompanying primary social goods to which these principles are applied. Secondly, I show that implementing a BI is an attractive alternative to implement the principle of fair equality of opportunity.

A.1. Rawlsian justice

Liberal-egalitarian theories of justice all try to realize an equal concern for all and non-discrimination among various conceptions of the good life. The most famous liberal-egalitarian theory of justice is 'justice as fairness' developed by Rawls (1971) in his book *A Theory of Justice* (hereafter abbreviated as TJ). Rawls's framework is used by Van der Veen (1991) and Van Parijs (1995) to argue for a BI. Rawlsian justice requires the distributive norm to be *maximin*: priority has to be given to the least advantaged in the distribution of real freedom. Real freedom (in Rawls's vocabulary: primary social goods) refers to the all-purpose means people need to pursue whatever conception of the good life they choose. It is important to note that real freedom can be measured either as the set of opportunities available or the amount of resources at one's disposal, but not as the outcomes or welfare reached by using these opportunities or resources. In this respect, it contrasts with utilitarianism which tries to maximize the sum of utilities over individuals.

That the ruling principle of the basic structure is to maximin the position of those with the least amount of real freedom is clear from Rawls's *general conception of justice*: 'All social values – liberty and opportunity, income and wealth, and the bases of self-respect – are to be distributed equally unless an unequal distribution of any or all of these goods is to the advantage of the least favored' (TJ, 303). The general conception of justice is not dependent on a particular conception of the good life (the so-called neutrality postulate). Using this list of primary social goods (liberties, opportunities, income, wealth, the bases of self-respect), the mutual priorities (or allowable tradeoffs) among them are not yet established. For instance, it may be possible to improve the income prospects of all by reducing some liberties or opportunities (for example the social bases of self-repect might be more favourable while receiving unconditionally a minimum income, as would be the case under a BI scheme than under a categorical benefit system, where the right to a social benefit is connected to all kinds of obligations and requirements). Therefore the mutual priorities and permitted tradeoffs between different social goods must be established. A partial answer is provided by the ranking of the following principles of justice, which can be regarded a special case of Rawls's more general conception of justice:

1) Each person is to have an equal right to the most extensive total system of equal basic liberties compatible with a similar system of liberty for all.
(2) Social and economic inequalities are to be arranged so that they are both:
(a) to the greatest benefit of the least advantaged, consistent with the just savings principle, and
(b) attached to offices and positions open to all under conditions of fair equality of opportunity (TJ, 302).

The principle of equal liberties (1) has a lexical priority above the principle of fair equality of opportunity (2b), which in turn has lexical priority above the difference principle (2a). The basic structure of society shaped according to these principles determines the distribution of primary social goods. Rawls (1982) describes these primary social goods as:

(a) First, the basic liberties as given by a list, for example: freedom of thought and liberty of conscience; freedom of association; and the freedom defined by the liberty and integrity of the person, as well as by the rule of law; and finally the political liberties;
(b) Second, freedom of movement and choice of occupation against a background of diverse opportunities;
(c) Third, powers and prerogatives of offices and positions of responsibility, particularly those in the main political and economic institutions;
(d) Fourth, income and wealth; and
(e) Finally, the social bases of self-respect.

The liberties under (a) are subject to the principle of liberty (1), and the goods under (b) guaranteeing free career choices are covered by the fair opportunity principle (2b). Powers and prerogatives of office (c) together with income and wealth (d) are subject to the difference principle (2a). Finally, the social bases of self-respect[30] (e), which Rawls considers as the most important primary social good, is covered by all three principles simultaneously.

The lexical priority rules of the above principles establish the ranking between liberties, opportunities and all the other primary social goods subject to the difference principle. However, it does not specify the relative weights to be attached to the goods of a given priority class. For instance, there is an index problem[31] for each of the goods (powers and prerogatives, income, wealth) subject to the difference principle (2a). As long as the list of primary goods and the weight attached to each good are unknown, we cannot determine to what extent a BI scheme meets the demands of justice as outlined by Rawls. If we would adopt the

---

[30] For an elaborate argument in favour of BI to protect self-respect, particularly among low productive workers, see Wolff (1998) and McKinnon (2003).
[31] This index problem was first mentioned by Arrow (1973, 254): '... so long as there is more than one primary good, there is an index-number problem in commensurating the different goods.'

rather ad hoc strategy for identifying the least advantaged by an index comprised of income only, then it is obvious that the BI scheme will be outperformed by a scheme which redistributes benefits selectively only to the poor. After all, a BI is paid out to all irrespective of income and wealth, while programmes targeted at the poor need only provide income support to those below the social minimum or the poverty level.

The general conception of justice stated above is nothing else than the difference principle applied to all primary social goods. This is why Rawls stipulates that '... in one form or another the difference principle is basic throughout' (TJ, 83). If the difference principle (or maximin criterion) applies to all social goods (liberties, opportunities, powers and prerogatives, income, wealth and the bases of self-respect), then in general terms (that is, disregarding the priority rules between the principles) Rawlsian justice does offer good points of departure for defending a BI. This is forcefully put by Van Parijs as follows:

> ... there is no doubt that an unconditional income confers upon the weakest more bargaining power in their dealings with both potential employers and the state, and hence a greater potential for availing themselves of powers and prerogatives, than a transfer contingent upon the beneficiary's availability for work and the satisfaction of a means test. Finally, Rawls mentions the social bases of self-respect, and there is again little doubt that a transfer system that is not targeted at those who have shown themselves 'inadequate' and involves less administrative control over its beneficiaries is far less likely to stigmatize them, humiliate them, make them ashamed of themselves, or undermine their self-respect (Van Parijs 1991, 105).

The idea that a BI strengthens the terms of negotiation of individuals with employers was the subject of this chapter. Closely linked to that is the notion of fair equality of opportunity.

A.2. The principle of fair opportunity[32]

What is meant by fair equality of opportunity, and what role has BI in this respect? The lexical priority rules imply that before the difference principle comes onto the scene, there should be equal liberties and fair equality of opportunity. The basic liberties do not change if current social security is replaced by a BI scheme. One interpretation of fair opportunities is that there must be fair competition for offices and positions. Fair competition, as in sports, requires only *formal* equality of opportunity, e.g. that no one is excluded to compete because of race, sex or religion. To put fair opportunities on a par with formal equality of opportunities is certainly not 'Rawlsian enough'. The other extreme, to put fair opportunities on a par with *strict* equality of opportunity, is neither very plausible because '... then fair equality of opportunity could require that all available resources be used to maximise the opportunities of those with the least opportunities. Creating opportunities for the least advantaged would take precedence over providing them with anything else' (Byrne 1993, 18). Indeed, there is no limit to policies (education, affirmative action,

---

[32] This section is for a large extent based on Byrne (1993, chapter 1).

job sharing programmes, etc.), and to the resources required to pursue these policies, to attain strict equality of opportunity. This would require that each person, no matter how untalented for a particular job, has the same chance to get it as anyone else.

In between these two extremes, an interpretation is needed which allows the principle of fair equality of opportunity to comply with the general conception of justice. According to this interpretation, *fair equality* refers to the opportunities open to the least advantaged relative to the more advantaged: equality is the baseline, and inequalities are only fair if they enhance the set of opportunities open to the least advantaged.[33] Recapitulating that the principle of fair equality of opportunity refers to the primary social good of free career choices, it is important to note that the subclause prescribes that these are made 'against a background of diverse opportunities' (see part (b) of the list of primary social goods above). This implies that agents are able to effectively choose from a sufficiently broad range of activities. Seen from this perspective, the involuntary (long-term) unemployed are the least advantaged because they are barred from the rewards of paid employment. Byrne argues further that BI can also be seen as a kind of wealth received.[34] To be sure, a BI does offer more than only formal equality of opportunity, but less than strict equality of opportunity. I do not claim that a substantial BI is the only policy which complies with fair equality of opportunity, but certainly it allows those with the least opportunities to price themselves into jobs (by using the BI as a wage subsidy), or to start their own business (by using the BI as a profit subsidy). It allows those who want to change career to opt out of paid employment for a while and to return to the educational system (by bearing the major part of the finance cost of schooling) and it allows those with a strong vocation unrecognized by the market to stick to it. Byrne (*ibid.*, 19) concludes that:

> Free choice of occupation implies that people should not have to work in any job that they would not freely choose and further that they should be free to change jobs if they so wish. The best way to bring this about is to grant a basic income to all of a sufficient size that people would no longer have to take the first job that comes along. The fear of being without a wage for a time would no longer discourage people from changing jobs or career direction. Arguably a state could devote all its resources into trying to create the 'background of diverse opportunities' referred to in the list of primary goods, against which people would choose their occupations. This would hardly constitute the best use of resources. However the introduction of a basic income, sufficiently high to constitute a living wage, would open up a multitude of diverse opportunities for all. People would have the freedom to engage in activities not presently recognised or rewarded by 'the market'. This would maximise the opportunities available to citizens beyond what any state sector, however large, could ever hope to. Basic income would guarantee the maximum occupational freedom and opportunity.

---

[33] In the section 'Further Cases of Priority' Rawls stipulates that 'an inequality of opportunity must enhance the opportunities of those with the lesser opportunity' (TJ, 303). That is to say, fair equality of opportunity is the difference principle operated on the good of opportunities.

[34] '... few writers ever really distinguish income from wealth... a basic income can be looked on as being similar to the income received by the rich from their fortunes... From this perspective we can see the "wealth" part of Rawls's fourth primary good as demanding that citizens get some kind of unconditional income' (*ibid.*, 26).

This statement duly describes the potential of BI to provide a broad set of background opportunities which goes even beyond that of paid work alone. A substantial BI facilitates much more than sheer formal equality of opportunity in both educational and labour market career choices, even when there are huge differences in social background. I therefore feel confident that the burden of proof of designing an alternative, which provides at least as good background opportunities for the least advantaged as they have under a BI scheme, rests on the opponents of BI. There are at least some reasons, although not conclusive, why a substantial BI and fair opportunities can go hand in hand.

To conclude this section I want to make a brief remark about another characteristic of a BI scheme conducive to effectuate Rawlsian justice. Under a BI scheme there is a transfer branch which hands out a BI to everyone, while minimum wages and the poverty trap are eliminated. Social security up to the social minimum is thus organized outside the labour market and there is less scope for the government to interfere on the labour market by all kinds of measures. The task of the transfer branch of the government according to justice as fairness is limited to effectively guarantee a social minimum.[35] This is in accordance with Rawls (TJ, 277) when he says that '... once a suitable minimum is provided by transfers, it may be perfectly fair that the rest of total income be settled by the price mechanism. Moreover, this way of dealing with claims of need would appear to be more effective than trying to regulate minimum wage standards and the like. It is better to assign to each branch only such tasks as are compatible with one another. Since the market is not suited to answer the claims of need, these should be met by a separate arrangement.'

---

[35] '... the government guarantees a social minimum either by family allowances and special payments for sickness and unemployment, or more systematically by such devices as a graded income supplement (a so-called negative income tax)' (TJ, 275).

CHAPTER 3

BASIC INCOME AND UNEMPLOYMENT[1]

1. INTRODUCTION

The debate on basic income (henceforth BI) seems to be contra-cyclical. The underlying reason for the coming and going of BI in and out of the picture, is the relationship between the welfare state and unemployment. In periods of recession, e.g. the 1930s, late 1970s and early 1980s, there are more discussions about BI than in periods of economic prosperity (e.g. during the so-called Golden Age of capitalism in the 1950s and 1960s). If unemployment is high and to a large extent involuntary, the policy to push the unemployed to accept (non-existent) jobs or to curtail social benefits becomes highly controversial, and as a corollary, forms of BI (or negative income tax) become more fashionable. Unemployment, in particularly large scale and long lasting unemployment, can safely be considered as one of the greatest problems for social policy makers. In the literature concerning the labour market and social policy, one not only finds theories that try to explain un-employment, but also many proposals to reduce or eliminate unemployment. These vary from piece-meal social-engineering approaches (e.g., cutting back social benefit and minimum wage levels, providing wage subsidies to low-waged labour), workfare-oriented approaches, to proposals which envisage an entirely different institutional framework of the labour market (e.g., Weitzman's (1984) *Share Economy*).

Whereas social policy makers and politicians must face the problems of the day, theorists have the prerogative and opportunity to indulge in fantasies and utopias. Behind their desk they can construct imaginary societies with institutions we never had the opportunity to experience in the real world. The BI proposal is such an imaginary construct. It would mean a major break with the means-tested and work-related social security systems now in force in most Western-European countries.[2] Some even speak of a new paradigm of social security based on the notion of fairness, beyond the present welfare state based on the principles of (Bismarckian) insurance and (Beveridgean) solidarity.[3] According to Goodin (2001), a BI would fit

---

[1] An earlier and more technical version of this chapter was published in *Recherches Economiques de Louvain (Louvain Economic Review)* 70 (2), 2004 under the title 'Basic Income, Unemployment and Job Scarcity'.

[2] For a recent overview of the potential of BI to address the new social question (the division between job holders and non-job holders) pitted against the claims of competing policy instruments (notably wage subsidies), and of the political chances of BI in various European countries, see Van der Veen and Groot (2000a,b).

[3] See Van Parijs (2000), which uses the term Painean justice to characterize unconditional basic social security provided by a BI.

in nicely into a post-productivist welfare state.[4] In the literature about BI, the link between BI (or the equivalent negative income tax) and unemployment is not scrutinized, despite the strong correlation between the intensity of the debate on BI as an alternative to the present, conditional, scheme of social security and the unemployment rate.[5]

The main aim of this chapter is to make plausible the claim that the case for a BI is stronger, and that the level of BI should be higher, the higher the level of structural unemployment. Structural unemployment is taken here in a broad sense: it refers to the shortage of jobs vis-à-vis the labour force, or more precisely, vis-à-vis total labour supply (therefore, as indicated in the title, under-employment is a better term in this respect than unemployment). For instance, even if the official unemployment rate would be close to zero, there might still be a considerable shortage of jobs, manifested by a large number of (notably female) persons, not entitled to social benefits, but willing to do paid work. To highlight the link between BI and unemployment, the BI scheme will be compared with the Labour Rights scheme as devised by Hamminga (1992; 1995), in which shortage of jobs is the point of departure motivating the whole exercize.

This chapter is organized as follows. Sections 2 to 4 discuss the equivalence between a Labour Right and BI system. Section 2 briefly outlines Hamminga's thought experiment, and in section 3 a formal model of the Labour Rights scheme is presented. Section 4 deals with the BI scheme. In section 5 the insights obtained from the analysis are used to critically examine welfare policy during economic up- and downturns. Section 6 evaluates the force of popular objections against BI, that of parasitism and of exploitation of hard-working citizens. In this context the latter objection is more relevant, because the framework laid out in this chapter offers an exquisite opportunity to test the exploitation objection by relating it to the insights gained from the comparison of the BI and Labour Rights systems. Section 7 deals with the issue whether or not the inclusion of employment rents warrants the choice for the maximum sustainable level of BI. The final section summarizes and concludes.

---

[4] Goodin (2000, 15-16) describes a post-productivist welfare state as follows: 'Under a post-productivist welfare regime, as under a social democratic one, people's welfare entitlements would be strictly independent of their participation in paid labour. But unlike social democrats, post-productivists would make no extraordinary efforts to get people into work alongside that. Post-productivists would take a relatively relaxed attitude to relatively large numbers of people drawing welfare cheques rather than pay cheques for relatively protracted periods. Of course, not everyone could do so. Like everyone else, post-productivists need enough people to work in the productive sectors of the economy to finance public transfers to those not. Post-productivists are not anti-productivist. Their point is simply that economic productivity can be sustained at moderately high levels on the basis of far less than full employment, full-time for absolutely everyone of working age. Post-productivists see this as a matter of social choice, collectively opting for a more relaxed life... Economic productivity can be sustained perfectly well without trying to put everyone into full-time work.'
[5] For this correlation, see Groot and Van der Veen (2000b, 197-9).

2. HAMMINGA'S THOUGHT EXPERIMENT

In a provoking thought experiment Hamminga considers a country Eu in which there are more workers than there are jobs available, but where everyone is given an equal and tradable entitlement to these scarce job assets.[6] Thus, the basic idea behind it is simply that a fair way of dealing with a shortage of jobs is to give everyone an *equal* and *tradable* share of Labour Rights. Doing this avoids a first-come, first-served appropriation of jobs (and assigning a social benefit to those who lose out), as well as make redundant measures like collective working time reduction to spread jobs over more workers. A market of tradable Labour Rights will have the nice result that workaholics will buy the Labour Rights of those with a low propensity to work, against an equilibrium price at which there are no non-workers who prefer to work, nor workers who prefer not to work and to cash in the market value of their share of Labour Rights. Consequently, in Eu it makes no sense to moralize about the level of the unemployment benefit, or whether or not able-bodied persons have a duty to work.[7] The main advantage of this system is that we get rid of both *in*voluntary unemployment and *in*voluntary employment. To see this consider Table 1, which classifies the labour force under the present conditional system into four categories, according to whether one is, voluntary or involuntary, employed or unemployed.

Table 1. Classification of the labour force according to labour market and motivational status under conditional social security (Hamminga 1995, 27, Table I: A Typology of Employment, adjusted).

Labour force	Voluntary	Involuntary
Employed	A	B
Unemployed	D	C

Given the levels of the unemployment or social assistance benefit and wages, those in B want to be in D, while those in position C want to be in A. In the Labour Right system it is only possible to be in A or D, because members of C will buy the Labour Rights (and thus take over the jobs) from B and become A's while those formerly in B move to category D. In Eu we have therefore only voluntary employment and voluntary unemployment, where the level of employment is determined by the number of jobs available, while the level of the unemployment benefit is determined by the equilibrium price of Labour Rights. The different

---

[6] In a country Eu there are five million able-bodied adult citizens, but only four million full-time jobs. The Eu-government gives everyone four Labour Rights, but to occupy a job one need to return five Labour Rights to the government. In total 20 million Labour Rights are issued by the government and also 20 million are needed for the four million jobs available. The one million people choosing to be unemployed, that is, those with a high preference for leisure, sell their Labour Rights to those who prefer, at the prevailing market price of Labour Rights, to work.

[7] 'In Eu, there's no discussion of whether people ought to work. It is not a matter of morals or politics or ethics. Jobs are like cars and concerts. Opting for employment is a matter of taste and your own preferred way of enjoying life' (Hamminga 1995, 26).

manners of how members of group D are treated in both systems is strikingly described by Hamminga as follows:

> Here, they are often distrusted, and we tend to feel no 'responsibility' for them: we do not want to feed them out of 'our' income for which we have the decency to work. D members are also a very useful symbol and instrument of demagogues who argue in favour of reducing unemployment benefits... In Eu, their presence is highly appreciated. They are the hard core of the supply side of the market in Labour Rights: they help keep down the price (*ibid.*, 32).

Contrary to the present system of social security, the preferences of all are reflected in the market price of Labour Rights. To sum up:

> In Eu's free market, all these preferences (LG: of those originally positioned in A, B, C or D) would be reflected in prices for which Labour Rights are bought and sold. In our world, it is otherwise. We moralize, making it a matter of politics and collective compromise. We concoct complicated criteria for deciding which inactive able-bodied adult citizens may receive a benefit, for determining the rate of the benefit, for deciding who is going to pay for it and how much. To Eunians, we look like communists deliberately organizing market failures for the sake of nineteenth-century morals (*ibid.*, 27).

The vantage point of the thought experiment is that it offers a very instructive device which compels us to (re)think how we treat the unemployed and what role the distinction between involuntary and voluntary unemployed (or 'deserving' and 'undeserving') plays. The experiment also shows that if unemployment is structural so that we have to face a long-lasting scarcity of jobs, the sacrifice of the workers (the price they have to pay for additional Labour Rights to secure a full-time job, and in this way pay for the unemployment benefits) mirrors the sacrifice of the unemployed to give up their right to work.[8] In this context, Goodin (see Schmidtz and Goodin 1998, 188fn) rightly notes that 'What they are doing for us is occupying slots among the unemployed that someone has to occupy, in an economy with any appreciable level of structural unemployment; and unemployment benefits can be conceived as payments to them for that service.' The 'taxes' paid by the workers to finance social security is not because of solidarity with the unemployed, but reflects a scarcity price which each one choosing to work pays because of self-interest. The solidarity of this system is so to speak a side-effect of the pursuit of self-interest by workers and non-workers alike. Despite these advantages in terms of fairness and efficiency, it is perhaps fair to say that in real world politics a full-fledged Labour Rights system is far less likely to be considered as a proposal worth to be taken seriously as an alternative to present social security than a BI or negative income tax (NIT) scheme. However, it will be shown that a Labour Right scheme is fully equivalent with a BI scheme, and therefore the advantages of the former, listed above, apply as well to the latter scheme.

---

[8] It cannot be ruled out that the unemployment benefit will even be higher than the net wage of those with a low productivity but a high preference to work, or even higher than the average net wage – this would be the case if to work is on balance not a sacrifice at all.

3. THE LABOUR RIGHTS SCHEME

3.1. Uniform productivity levels

In the experiment described by Hamminga (1995) a homogeneous population (that is, with equal earning capacities) is assumed, with only a choice between full-time work or full-time leisure. In what follows, these simplifying assumptions will be lifted sequentially.

Consider an economy in which the maximum feasible employment level, expressed as a proportion of the labour force, is f (so f is the ratio between the number of full-time equivalent jobs[9] and the total labour force), with $f < 1$. The government, acknowledging this shortage of jobs, decides to give every member of the potential labour force p Labour Rights, with $0 < p \leq f < 1$. If someone wants to work full-time, she has to buy an additional amount or share of $(1-p)$ Labour Rights on top of the p Labour Rights which she holds already in possession. Any worker can thus freely decide how many hours of a standard work week to work by selling or buying the required amount of Labour Rights. Since we first maintain the assumption of equal talents (and equal wages), the price to be paid for additional Labour Rights (p_c) can be interpreted as a proportional tax on labour income (w). Workers differ with respect to their propensity to work: those with $e = 0$ are the most work-averse (preferring a leisure-oriented life), while those with $e = 1$ are the workaholics.[10] Since each member of the workforce receives p Labour Rights at the beginning, and given a uniform, equilibrium, market price of p_c, the income resulting from selling all one's Labour Rights (so not participating on the labour market) is equal to pp_c. Likewise, someone who just decides to perform paid work during a fraction p of a full-time work week does not have to buy additional Labour Rights, nor has to sell any Labour Rights, and receives income wp. Those who wish to work longer than p (times the number of hours of a standard full-time work week) have to buy additional Labour Rights. For example, a full-time worker has to buy $(1-p)$ additional Labour Rights on top of the p Labour Rights which this worker already holds in possession.

From the Appendix, in which the full model is explained, the following shorthand expression for p_c, the equilibrium price to be paid for a coupon of Labour Rights, is reproduced:

$$p_c = \frac{\bar{e}w - p}{\bar{e}} \tag{4}$$

Fixing the uniform wage rate (w) at unity, and given that shortage of jobs $(p \leq f \leq 1)$ is the point of departure motivating the whole exercize, the price of Labour Rights is

---

[9] Two part-time jobs of half a standard work week count as equivalent to one full-time job.
[10] This way to model preferences for work and leisure, the utility function, and some other assumptions are the same as in Chapter 3 of the dissertation of Vandenbroucke (2001).

higher, the higher is the average propensity to work $\overline{e}$ and the lower the chosen employment level p. (If everyone's share of Labour Rights is higher than $\overline{e}$, then there is excess supply of Labour Rights, which causes the price of a Labour Right to dwindle to zero.[11]) The logic behind the Labour Right market is that the net money income of a full-time worker will be lower, the lower the pre-established employment share p (which makes jobs more scarce) and the higher the average preference to work $\overline{e}$ (indicating that many others want to work hard, which is to the advantage of those who have a low preference for working). Figure 1 depicts the relation between non-labour income (pp_c) and the price of labour rights p_c, both on the vertical axis, against the level of the employment rate p, on the horizontal axis.

Figure 1. The price of Labour Rights (p_c) and the level of the unemployment benefit (pp_c) as a function of the participation rate p

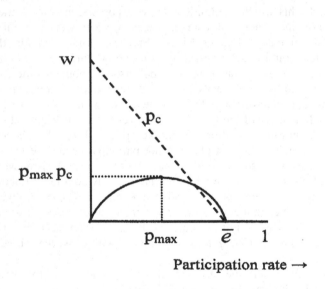

Participation rate $\rightarrow$

The interrupted line shows that p_c is a declining function of p (and equal to zero if p is equal to $\overline{e}$): the higher p, the higher the volume of Labour Rights circulating in the economy as a whole, hence the lower will be the market price p_c for a unit of Labour Rights. Non-labour income, which can be seen as a kind of unemployment benefit (although part of it can also be cashed in if one is working part-time (as long as labour supply is below the fraction p of a full-time job), is represented by the solid line and shows a hump-shaped curve. There is a unique vale for p, say p_{max}, at

---

[11] If $\overline{e} < p$, the price of Labour Rights would become negative only if everyone is *enforced* to excercize the Labour Rights, that is either to work more than one wants to or to find someone who is prepared to take over Labour Rights.

which non-labour income is at maximum. For p less than p_{max}, the price of Labour Rights is high because jobs are scarce, but anyone choosing not to work has only a small share p of Labour Rights to sell to those who want to work more than their modest initial granted share. One can also take it in this way, that for a low p there are relatively few workers doing paid work and many abstaining from paid work, which implies a low average income for both workers and non-workers. If p is higher than p_{max}, e.g., close to $\overline{e}$, then jobs are not so scarce and the price for a Labour Rights is consequently low. The fact that non-workers can sell a larger share of Labour Rights compared to the situation where p is equal to p_{max} does not compensate the lower equilibrium price of Labour Rights.

To summarize, given the value of $\overline{e}$, we can determine the share of Labour Rights (p_{max}) to be distributed equally to all which maximizes the non-labour unemployment income ($p_{max} \, p_c$). However, so far nothing is said about whether to choose the participation rate which maximizes the unemployment benefit is also just. It may well be that the maximum *feasible* overall employment level or participation rate (f) is far higher than the rather low participation rate p_{max} which maximizes the unemployment benefit. Note that the pre-established participation rate p determines how large a share of a full-time job right each member of the labour force receives. As said (see also Equation (4)), the market price p_c for a Labour Right approaches zero if p approaches $\overline{e}$. It is favourable to those with a high preference for work if the government sets the overall participation rate p close to the maximum feasible rate, or in the limit to 1, because in that situation they only have to buy a small additional share ($1-p$) Labour Rights and, moreover, against a cheap price, if they decide to work full-time. The hard core of the voluntary unemployed however would favour a participation rate which is close to the rate p_{max} which maximizes their unemployment benefit. The question then arises which rate the government must choose.[12] The answer to this question is rather straightforward. Instead of artificially constraining the number of jobs in order to maximize the unemployment benefit, or to artificially expand the number of jobs to the advantage of the workaholics, the government acts in a perfectly equitable way if it sets p equal to the feasible employment level f, whatever its level turns out to be. If f is far higher than p_{max}, we feel sorry for the leisure-lovers, but they have no valid complaint. Equally, if f is equal to or close to p_{max}, we feel sorry for the workaholics, but they have no valid complaint either. In both cases, everyone has equal access to jobs, and given equal productive talents, everyone has also equal opportunities to convert leisure into money income. There is no injustice and the problem of shortage of jobs is handled in an equitable way. This issue is taken up again in section 4 when discussing the equivalent BI scheme.

---

[12] This problem turns out to be quite tricky, not only because of the opposing interests of workaholics and leisure-lovers. For instance, the maximum feasible rate is higher if the conditions to classify jobs offered by employers as proper jobs, are loose: do we count the jobs offered by an employer who wants to start a peep-show as proper jobs? Again, it is in the interest of those with a low attachment to paid work to have rather strict criteria (which brings the pre-established rate closer to p_{max}), while it is in the interests of full-time workers and employers that rather loose criteria are used.

3.2. Non-uniform productivity

If the assumption of equal earning capacities (a uniform w) is removed, we encounter the problem that in the original set up of the experiment the price to be paid for additional shares of Labour Rights is essentially a tax imposed on workers proportional to working time, but not proportional to their earning capacity. It thereby violates an important principle of tax equity, namely that everyone pays according to one's means. For someone with a rather low earning capacity, but a high preference to work, the tax in the form of the uniform price of additional Labour Rights may constitute a serious obstacle to choose the option of full-time work, whereas for someone with an extremely high earning capacity, it is not more than a tip. In order to circumvent this problem, just assume that all workers pay a tax proportional to their earning capacity w and their labour supply (so the tax liability is twL), and that the proceeds of this tax is distributed equally to all. Whether one is a net tax recipient or contributor depends on whether the (earning power related) tax liability (twL) is smaller or greater than the uniform benefits received (pp_c). Also in this state of affairs the free trade in Labour Rights will result in an market clearing equilibrium price for Labour Rights. Both the tax rate (see Equation (10) in the Appendix) and the price of Labour Rights p_c are fully determined by the chosen level of p, the average propensity to work ($\bar{e}$) and the average productivity level ($\bar{w}$). The again reproduced summarizing formula for the level of non-employment income is:

$$p_c = \frac{(\overline{ew} - p)/\bar{e}}{(\overline{w}^2 + \sigma^2)/\overline{w}^2} \tag{11}$$

As with uniform productivity (see Equation (4) above), the price of Labour Rights is higher, the higher is the average propensity to work $\bar{e}$ and the lower the chosen employment level p. However, the difference is that with non-uniform productivity also the variance in productivity levels is of importance. Note that if $\overline{w}$ is set to unity (or w), and consequently the variance is zero, Equation (11) is identical to Equation (4). Here we encounter exactly the same problem as before, namely the choice for the level of p. For those who want to live merely from the proceeds of selling Labour Rights, it would be optimal to choose $p_{max} = \overline{ew}/2$, the same as we found in the case of a homogeneous workforce, except that the uniform wage rate w is now replaced by the average productivity $\overline{w}$, where p_{max} corresponds to a tax rate equal to 50%. However, it is far from obvious that an egalitarian government, faced with the problem of scarcity of jobs, should aim at maximizing the income of those who want to indulge in leisure. Instead, the maximin objective requires to maximize the income of the group of least advantaged (those with the lowest productivity level w_L), and to weight preferences impartially. Doing this requires to set the tax rate below 50% if $w_L > 0$: the higher the lowest productivity level w_L, the lower the required tax rate will be, hence the less redistribution. In the limit, as w_L approaches $\overline{w}$, so that c^2 gets smaller and smaller, the tax rate tends to zero: if all have more or less the same productivity w, and the same access to jobs, then there is no rationale for taxation and redistribution. At the other extreme, if w_L is zero, the least productive are not able to capture any labour income, so the best thing for them to do is to sell their Labour Rights, and the best thing for an egalitarian government to do for them is to set the tax rate at the Laffer turning point which maximizes non-labour income. The higher the value of w_L, the lower the required tax rate to achieve the maximin objective and therefore the closer the employment level p approaches the maximum feasible level f.

Admittedly, to ration artificially the employment level below the maximum feasible level is probably an outcome hard to swallow, especially if we bear in mind that the whole exercise departs from the presumption that scarcity of jobs is the main problem. For several reasons this outcome is nonetheless acceptable. Firstly, suppose that instead of rationing the employment level the government would allow everyone to work as much, but not more, as what corresponds to everyone's legitimate share of Labour Rights. This would amount to giving each member of the labour force an *untradable* share of Labour Rights equal to the feasible rate f. This measure is equivalent with a compulsory working time reduction of all full-time jobs of $(1-f)$ percent. It is clear that there will be ample room for Pareto-improvements if Labour Rights would be tradable: those wanting to work longer than f hours of a standard work week would be eager to pay a price for additional Labour Rights from those who want to work less than f hours. It is the tradability of Labour Rights which ensures that all possible Pareto-improvements can be realized. Secondly, the fact that $p < f$ only results because the government wants to maximize the income of the least advantaged. In a sense, then, the divergence between f and p can be interpreted as the cost in terms of employment of providing social security, with f the employment level in an economy without any or only minimal social security. One more reason why the result of putting a ceiling to the employment level should not bother us has to do with the fact that it is only a thought experiment. As will be shown in the next section, the problem does not show up in this guise under an equivalent BI scheme.

4. THE EQUIVALENT BASIC INCOME SCHEME

4.1. Uniform productivity levels

A flat tax BI scheme (or a single tiered NIT scheme) is fully characterized by two parameters: the level of the BI (or the NIT income guarantee) and the tax rate.[13] In the Appendix, the set of equations describing a BI scheme are presented, where apostrophes are used to indicate the equivalence with the model with Labour Rights. Again, only the most salient equation is reproduced, which is the level of basic income (B):

---

[13] See Atkinson (1995a) for an extensive treatment of the BI/flat tax proposal.

$$B = \overline{e}[t(1-t)w^2] \qquad\qquad (4')$$

It can easily be checked that differentiating Equation (4') with respect to the tax rate
(*t*) gives the *B*-maximizing tax rate $t_{max} = \frac{1}{2}$. The relation between *B*, total labour
supply (*L*) and the tax rate is shown in Figure 2.

Figure 2. The level of basic income B and labour supply L as a function of the tax rate t

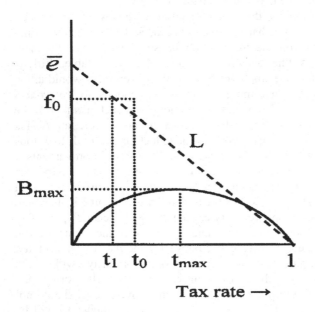

In the Appendix it is shown that *B* is equal to pp_c. An easy check is to substitute pp_c
for *B* and p_c for *tw* in Equation (4'), which gives the expression (see Equation (4)
above) for non-work income under a Labour Rights scheme, which establishes the
formal equivalence between both schemes.

The only difference between both schemes is this: in the Labour Rights scheme
the government must choose beforehand the participation rate *p* (which determines
how large a share of a tradable job rights everyone is granted), which in turn
determines, given $\overline{e}$, the value of p_c (the 'tax rate') and hence the level of the
unemployment benefit; under the BI scheme, the government has to decide
beforehand which tax rate to impose, which as a matter of fact simultaneously
determines the overall participation rate (or total labour supply) and the level of *B*.
In the former scheme, the participation rate *p* is taken as an exogenous variable
while the 'tax rate' p_c and the level of the unemployment benefit are endogenous; in
the latter scheme the tax rate is exogenous, while the participation rate and the level
of *B* are endogenous.

The right procedure for the government to follow is to choose a tax rate t_0, and to check whether there are still vacancies left (so $f_0 > L$ at $t = t_0$, see Figure 2). If so, the tax level and level of B are too high, causing labour supply to be insufficient to fill all jobs available. Next, lower the tax rate (to t_1) as well as B, so inducing a higher labour supply, until all job vacancies are disappeared. If, on the other hand, at $t = t_0$, there are people searching in vain for a job (so $f_0 < L$ at $t = t_0$), then labour supply is too high compared to the number of jobs. To reach equilibrium on the labour market, the government can raise the level of t as well as B, inducing a decline in labour supply, until total labour supply matches labour demand. In conformity with the Labour Rights system (see Table 1 in section 2), at the market clearing levels of t and B, there is now only voluntary unemployment and voluntary employment.

4.2. Non-uniform productivity

The exercise with unequal talents is straightforward and results in the following expression for B:

$$B = \overline{e}t\,(1 - t)(\overline{w}^2 + \sigma^2),\qquad(11')$$

so the tax rate that maximizes B is $t_{max} = \frac{1}{2}$. Maximizing average income of the group with lowest productivity yields that the tax rate will be lower, the higher the lowest productivity level w_L in exactly the same functional relationship as found in the Labour Rights scheme.

5. WELFARE POLICY AND ECONOMIC UP- AND DOWNTURNS

Now that we have some insight in the functioning of a Labour Rights and the equivalent BI scheme, it is interesting to see how these schemes react to exogenous economic shocks compared to how welfare policy usually accommodates changes in the economic environment. It is taken for granted that labour demand is always the short side of the labour market (for most European countries, this is a plausible assumption for already three decades as far as low wage jobs is concerned). Under a Labour Rights scheme, a positive shock to the economy would allow the government to set the participation rate p at a higher value, leading to a decline in the level of the unemployment benefit[14] (provided $p > p_{max}$). In terms of Figure 1, the economy moves further to the right, away from p_{max}. Under a BI scheme, a positive shock (an increase in labour demand at prevailing wages) effects a decline in the level of B. In terms of Figure 2, the tax rate must be set at a lower level to foster labour supply in order to absorb higher labour demand, so the economy moves to the left, e.g. from t_0 to t_1, away from t_{max}. This may come as a surprise, but it is in line with the central assertion of the analysis that the legitimate level of B is lower, the

---

[14] Which is of course necessary to stimulate labour supply, or in terms of Table 1, to give some members of group D (the voluntary unemployed) the incentive to become members of group A (the voluntary employed).

lower the level of unemployment due to shortage of jobs. The limiting case is where there is no shortage of jobs at all, in which case the legitimate level of B is nil under the assumption of equal talents.[15] Under the existent conditional welfare scheme, a positive shock also reduces the number of unemployed. The difference is that a lower unemployment rate removes somewhat the pressure to lower social benefits: if anything, the higher tax revenues and lower social security outlays allows in principle to raise the social benefit and to lower tax rates.

In an economic downturn, this picture is reversed. Under a Labour Rights system, the participation rate declines and unemployment benefits increase (which is necessary to induce more workers (members of group A according to Table 1) to voluntarily abstain from paid work and to become voluntarily unemployed (and become members of D). The same is true under the equivalent BI scheme: a higher tax rate is required to reduce total labour supply until it equals labour demand (which as a result of the economic downturn decreases). Under both schemes, the economy moves closer to p_{max} (again, provided the participation rate before the shock occurs was higher than p_{max}). The usual welfare policy, the best remedy under the existent social security scheme so to speak, to overcome an economic downturn and the problems this poses for the welfare state is the opposite move, namely to curtail social benefits. Why?

The first reason one can think of is that lower social benefits effects a higher labour supply because it is usually assumed that the latter varies inversely with the replacement ratio. However, even if all unemployment before the shock occurs was voluntary, this measure is unnecessary because some workers will lose their job and become involuntary unemployed. An economic downturn will thus inevitably, even if social benefit levels are maintained, create involuntary unemployment (exerting a downward pressure on wages) which suggests that it is unnecessary to further stimulate (excess) labour supply and downward pressure on wages through cutting social benefits. The second reason is that higher unemployment exerts an upward pressure on tax rates because of increasing social security outlays, which in turn raises gross wages (and hence would even aggravate the decline in labour demand, leading to a vicious circle). This kind of reasoning implicitly assumes that those who remain at work do not accept a decline in net wages despite the economic downturn. If this is true (and this is very plausible, since insiders have sufficient power and job security to refuse net wage cuts), it means that only the unemployed must carry the burden of overcoming the economic downturn: lower social benefits allows lower tax rates and, given the level of net wages, lower gross wage costs for employers. To be sure, the economic downturn can then only be overcome when the higher labour supply and lower tax rates indeed lead to lower gross wage levels and a corresponding higher labour demand. To summarize, the difference in adjustment to a negative shock between the Labour Rights and BI scheme on the one hand and the prevailing welfare policy under the existent scheme on the other is that the former schemes just try to make all (un)employment voluntary (which requires higher

---

[15] Or positive in the case of unequal talents, not because of shortage of jobs, but merely because the government uses the redistributing instrument of B to maximize the average income of the least advantaged.

unemployment benefits and lower net wages for workers), while the latter always tries to return to a state of full employment (whether voluntary or involuntary), if necessary by means of cutting social benefits.[16] Under the Labour Rights and BI schemes, at each level of unemployment the unemployment benefit or BI is set at that level at which all (un)employed are in that position voluntarily; the higher the level of structural un(der)employment, the higher the required level of income for the unemployed to induce sufficient workers to choose for unemployment.

6. PARASITISM AND EXPLOITATION

The exploitation objection against BI, formulated by Elster (1986, 719), runs as follows: 'Most workers would, correctly in my opinion, see the proposal as a recipe for the exploitation of the industrious by the lazy.' The exploitation objection must be distinguished from the reciprocity-based parasitism-objection, which says 'no benefit without work' (see e.g. White 1997). In the latter, the reciprocity-based duty to work (those who can work should not be entitled to social benefits) is at stake, while in the former the lower standard of living of workers in virtue of supporting the parasites is involved. However, if there is a shortage of jobs, and no shortage of qualified workers prepared to occupy these jobs, it makes less sense to force unemployed workers to meet the demands of reciprocity. Probably White (*ibid.*, 82) might accord with this view since he writes: 'Since it is only fair to insist on satisfaction of the reciprocity principle if there are sufficient opportunities for citizens to do so, it would arguably become unfair to continue to affirm the principle were we unable to return to full employment (in an appropriately modernised sense). The case for citizen's income (basic income) would then be correspondingly stronger.' If one nevertheless wants to hold fast to the demands of reciprocity, then the workaholic must give up part of his full-time job.

Needless to say, the framework expounded in the previous sections offers an excellent opportunity to evaluate the merits of the exploitation objection. As a first reply one might say that under a BI (or Labour Rights) scheme any worker has the opportunity to choose a fully leisurely life-style, so those workers who feel exploited can immediately switch to a situation with full-time leisure. This reply will not do, because it is just the moral rightness of providing unconditional income transfers which is at stake. Moreover, in the counterfactual case that indeed everyone would take this opportunity, there is no BI to dispense. The real interesting question is therefore to figure out what reasonable complaints workers may have when the BI turns out to be economically feasible, that is, at a level where enough persons want to convert leisure time into additional money income (on top of their BI) by means of paid labour (and so keeping that level of BI economically sustainable). Van der Veen (1991, 203) replies that '... disgruntled workers have no valid moral

---

[16] Note that the relevance of this conclusion, and also of this chapter, becomes much greater if we interpret the existing social benefits as conditional on means only, while the work-test is merely formal (to fulfil the formal duty to work, the unemployed only have to visit regularly the job centre to see that no job vacancies comes along). In that case, voluntary unemployment is, not in theory but in practice, tolerated.

complaints against the behaviour of their non-working fellows under a genuinely sustainable universal grant'. Now disgruntled workers are those who decide '... to stay in the workforce, even though they would rather quit, [is] explained by the moral motivation of not wanting to be parasitic' (*ibidem.*). In other words, these workers do have the same preferences regarding work and leisure as those who choose full-time leisure, except that they do not want to live off others labour. Hamminga's experiment is of great help here since under a conditional scheme there is no possibility to test how *sincere* the opinions of the disgruntled workers are, while the sincerity and the alleged ubiquity of these opinions among workers are put to a real test if Hamminga's proposal for freely tradable Labour Rights in the circumstance of scarcity of jobs is carried out. It might well turn out that the alleged disgruntled workers under a Labour Rights scheme would quickly choose the option to full-time leisure by selling their Labour Rights, if only because this is a perfectly legitimate move under that scheme, and therefore do not attach much importance to the notion that it is morally suspect to live off others labour (which, according to Hamminga, is the official work ethic uphold by the existing work- and means-tested scheme of social security).

Nonetheless, with an equal social minimum, the balancing budget tax rate of the BI scheme will be higher than the average tax rate of a conditional scheme with categorically targeted and means- and work-tested benefits. The question is then whether it is fair to tax workers with a high preference for work over leisure in order to finance a BI scheme.[17] To keep within the confounds of the analysis of the thought experiment elaborated in the previous sections, it is clear that the government must always aim at the maximum feasible employment level. The only exception to this rule is that following the maximin objective, that is to maximize the income of the least advantaged, might lead to a situation where redistribution coincides with a lower participation rate.

Returning to the Labour Rights scheme, if there is a scarcity of jobs due to structural un(der)employment, there is room for an unconditional income entitlement, even for those who voluntarily choose to be idle. However, to pitch the level of non-labour income against the highest durably sustainable level (financed by the single proportional tax rate which yields the highest tax revenues, hence the highest feasible unemployment benefit or BI) would be unjust to (and indeed exploiting) all those who want to work hard. The tax rate corresponding to the case where the government tries to maximize the income of the least advantaged, while taking into account the scarcity of jobs, need not be equal to, and probably will be far less than, the tax yield maximizing rate. This means that in these circumstances there exists a positive level of the unemployment benefit or BI, albeit less than the highest feasible level, which can withstand Elster's objection: it is not because they are industrious that they have to pay taxes, but because they want to appropriate

---

[17] Recall Elster's statement (1989, 215): 'People who chose to work for an income rather than to live in a commune on the universal grant would have to pay higher taxes in order to support those who took the other option. They would think, correctly in my opinion, that they were being exploited by the other group.'

more than their legitimate share of scarce job assets, and because of the required amount of redistribution to improve the position of the least advantaged.

7. (UN)EMPLOYMENT RENTS

Under the assumptions of scarcity of jobs and equal talents, the answer to the question 'Is the justified level of BI also the maximum feasible level of BI?' is negative. In general, the former is much lower than the latter. The maximum sustainable BI only depends on the average preference for work, whereas the justified level of BI depends on two parameters: the shortage of jobs (which determines the justified tax rate) and (indirectly) the average preference to work. The approach (equal talents, tradable job rights in the case of scarce jobs) adopted in this chapter is the same as the one used by Van Parijs (1995, chapter 4), but the conclusion is contrary to his claim that: 'At least in a context in which no account is taken of differences in skills, the maximin or leximin version of our generalized Dworkinian criterion recommends that wages should be taxed up to the point at which the tax yield, and hence the basic income financed by it, is maximized' (ibid., 116).

Obviously, Van Parijs must have included some elements which are overlooked in the thought experiment discussed above. The most important element is that in a non-Walrasian world labour markets do not clear, and wages are higher (e.g., due to minimum wages, union bargaining, insider-outsider effects and efficiency wages (ibid., 107)) than would be the case if the labour market invariably clears. The difference between these non-market clearing wage rate and the market clearing wage rate constitutes an employment rent, which, according to Van Parijs (ibid., 108), should be taxed in maximin fashion: '... room is made for a sizeable increase in the level of basic income that is warranted on real-libertarian grounds. It amounts to sharing among all the employment rents otherwise monopolized by those in employment. Those rents are given by the difference between the income (and other advantages) the employed derive from their job, and the (lower) income they would need to get if the market were to clear'. In the thought experiment and the equivalent BI model outlined above, nothing was said about whether the wage rate (w) was higher than or equal to a market clearing level. Still, even if it is higher (say, this higher non-market clearing level is pre-eminently the cause of the shortage of jobs), does this constitute an additional source to swell the level of a justified BI (perhaps close to the maximum sustainable level)? Consistency requires that not only the rents received by workers are taken into account, but also the unemployment rents enjoyed by unemployed workers. Analogous to Van Parijs's definition of employment rents, unemployment rents can be defined as the difference between the unemployment income (and other advantages) the unemployed derive from being unemployed and the (lower) income they would need to get if the market were to clear. Now certainly the unemployment benefit in a perfect Walrasian, thus clearing, labour market is nil. Therefore, the entire unemployment benefit (or BI) should count as an unemployment rent.

Suppose that both type of rents were not taken into account in the analysis above. What would happen if we include them in the tax base? Taxing both employment and unemployment rents alike, and distributing them equally among all, increases the level of a justified BI only if the (per capita value of) employment rents enjoyed by workers is higher than the (per capita value of) unemployment rents enjoyed by non-workers. However, the level of both the maximum sustainable and the justified unemployment benefit (or BI) is higher, the higher the level of the wage rate, and the higher the level of unemployment (see Equations (4) and (4'), and (11) and (11')) . In other words, how much workers are prepared to pay for additional Labour Rights (Equation (4)), or how much labour they want to supply given the tax rate and hence the level of the BI (Equations (4') or (11')), is dependent on the level of the wage rate. Consequently, the effect of wage rates exceeding the market clearing level, and hence the occurrence of employment rents due to non-market clearing, is already taken into account. If wages are above market clearing levels, and given proportional taxation of labour income, then the rent component of wages is already part of the tax base (in the model with non-uniform productivities, taxation is already set according to the demands of the maximin-objective). The unemployment rents, however, are not yet taken into account because the unemployment benefit (or BI) is untaxed. Taxing them and distributing the proceeds equally lowers the level of the unemployment benefit (or BI), since part of these tax revenues accrues to the workers. The existence of (un)employment rents due to inevitable unemployment is thus not a factor which can explain why the justified level of BI is far below the maximum sustainable level of BI.

Another element which Van Parijs advances is that even when talents or skills are equal and there is full employment, there can still be a variety of jobs, leading to job rents. Given the assumption of equal talents, only preferences (or personalities) of workers can determine the matching of workers to different jobs. It cannot be ruled out that some workers prefer a job occupied by another worker (so-called envy over job endowments), and this constitutes another form of employment rent (a job rent):

> ... as soon as there are several types of jobs, the existence of employment rents no longer needs to be coextensive with involuntary unemployment: there may be huge employment rents even if everyone has a job, because many people with lousy jobs may be willing and able to do other existing jobs far more attractive (financially or intrinsically) than theirs at the going wage. What is crucial to my argument is the existence of large employment rents, and not the fact that many people are without a job at all. The conclusion, therefore, fully applies to affluent countries, such as the United States, in which the rate of unemployment is comparatively low, just as much to Western Europe' (ibid., 109).

However, if justice requires taxation of job rents, than it also requires taxation of the symmetrical non-job rents (or leisure rents) enjoyed by the non-workers: the opportunities open to lead a meaningful or satisfying life (e.g., those with families to care for, extensive networks or interesting hobbies) may vary as much as jobs differ intrinsically, giving rise to envy over non-job endowments, not only among fellow unemployed persons, but also among workers. Analogous to the (un)employment rents, the existence of job and non-job rents due to varying jobs or opportunities open to (un)employed workers may make a difference for the level of the justified BI. Whether including these type of rents in the tax base changes the level of the unemployment benefit or BI again depends on the volume of these rents enjoyed by both groups of workers and non-workers. It is however implausible that the existence of these job and non-job rents warrants the implementation of the maximum feasible level of BI as a matter of justice. To be sure, the maximum feasible level of BI arises when the preferences for work of the labour force is fully 'exploited' in the sense that the interests of those with the lowest preferences to work are optimally served.

SUMMARY AND CONCLUSIONS

The analysis of this chapter is entirely concerned with scarce job assets. However, this methodology of granting equal and tradable rights can be transposed to other areas. For instance, the same approach can be adopted if we are concerned with a just distribution of pollution rights.

The aim of the analysis performed is to investigate whether there is a relation between un(der)employment and BI. The main question is what workers have to pay to appropriate (scarce) jobs assets. As a starting point it is assumed that for one reason or the other there is scarcity of jobs. Hamminga's proposal to deal with scarcity of jobs is to give each member of the labour force an equal and tradable right to these scarce job assets. It is shown that such a Labour Right system is equivalent with a BI scheme, with the only difference that in the former the level of employment (or participation rate) is exogenous and the 'tax rate' endogenous, whereas under the BI scheme it is the other way around. The equivalence consists in that the price of Labour Rights and the level of the unemployment benefit corresponds to the proportional tax rate and the level of BI respectively. In other words, the single proportional tax rate of a BI scheme can be considered as what workers have to pay to appropriate scarce job assets (and for redistributional objectives). Both schemes allow that some people voluntarily abstain from doing paid work, in return for a financial compensation, an unemployment benefit or a BI.

Starting from no scarcity of jobs at all, the level of the unemployment benefit or BI varies positively with the degree of scarcity of jobs until the maximum value is reached. The overall participation rate corresponding with the maximum value of the unemployment benefit or BI is rather low, certainly when compared with the average preference to do paid work. There is, however, no reason to choose for the maximum sustainable BI. Under equality of talents, the government can simply choose the level of employment which it deems feasible, and let the market in Labour Rights determine the level of the unemployment benefit. Under the BI scheme, the government can set the proportional tax rate (and the level of B) at the level at which the labour market clears.

The bottom line of the analysis is that the level of BI varies positively with the level of unemployment: more severe scarcity of jobs requires a higher tax rate (and hence a higher BI) to adjust the total labour supply downwards to the number of jobs available. The level of the BI also varies positively with the average preference to work. The logic behind this is that, given the tax rate, more people voluntarily choose to work if the average preference towards work is higher. So in a society with a strong work ethic it is possible to have a high BI as well as a high level of employment (that is, close to the maximum feasible employment level) in conjunction with a low rate.

APPENDIX

In this appendix the underlying model will be presented. Following the classification in the main text, a distinction is made between the Labour Rights scheme and the Basic Income scheme. In addition, within each scheme productivity levels can be uniform or non-uniform.

A1. The Labour Rights scheme, uniform productivity

Eunians face the following budget constraint:

$$Y = [w - p_c]L + p\, p_c \qquad L \in [0, 1], \qquad (1)$$

with
Y = net income;
w = the wage rate;
p_c = the equilibrium price for a unit of Labour Rights;
p = the initial share of Labour Rights;
L = labour supply.

The labour supply[18] L is equal to:

$$L = e\,[w - p_c] \qquad e \in [0,1], \qquad (2)$$

where the parameter e expresses the, individual specific, propensity to perform paid work. Note that this labour supply function has no income effect and a unitary elasticity with respect to the net wage. Empirical research invariably finds that male labour supply is rather inelastic (close to zero), while female labour supply is somewhat more elastic, but its elasticity also well below unity. One can easily adjust the labour supply function according to Equation (2) for varying elasticities, through specifying $L = e\,[w - p_c]^\varepsilon$, so that the elasticity is equal to ε. However, this would make the analysis much more complicated. The main thing to bear in mind is that the level of unemployment income (or BI) can be much higher with inelastic labour supply than the levels which follow from the analysis here. The equilibrium condition for the market of Labour Rights can be expressed as follows:[19]

---

[18] It can be easily derived that the labour supply function according to Equation (2) corresponds to the following utility function $U_e(Y, L) = Y - L^2/2e$, where the second term on the RHS represents the burden of work.

[19] Alternatively, one might specify this as either:

$$\int_e L f(e) d(e) = p \tag{3}$$

stating that total labour supply must equal the employment level chosen by the government ($f(e)$ is the density function of parameter e). Substitution of (2) into (3) and solving for p_c gives:

$$p_c = \frac{\bar{e}w - p}{\bar{e}}. \tag{4}$$

which is the equation that was reproduced in the main text. Non-labour income, that is money income resulting from selling one's initial share p of Labour Rights against the market price p_c, is maximized for that value of p for which:

$$\frac{\partial(p\,p_c)}{\partial p} = 0 \Rightarrow p_{max} = \frac{\bar{e}w}{2}. \tag{5}$$

where p_{max} denotes the participation rate which maximizes the unemployment benefit. Figure 1 in the main text shows the relationships between the unemployment benefit pp_c and the market clearing price of Labour Rights and different values of the pre-established participation rate p, with preferences ($\bar{e}$) held constant.

A2. The Labour Rights scheme, non-uniform productivity

With non-uniform productivity levels it is assumed that all workers pay a tax proportional to their earning capacity w and their labour supply (L). Accordingly, the budget constraint and labour supply function become:

$$Y = [(1 - t)w]L + p\,p_c \qquad w \in [w_L, 1], \tag{6}$$

and

$$L = e[(1 - t)w]. \tag{7}$$

$$(3') \int_0^{e_p} (p - L) f(e) d(e) = \int_{e_p}^1 (L - p) f(e) d(e) \text{ or } (3'') \int_e L\, p_c f(e) d(e) = pp_c.$$

According to (3'), the amount of Labour Rights sold (the LHS) must equal the amount of Labour Rights bought, where e_p is the propensity to work of someone who chooses to work exactly p. Equation (3'') is the usual balanced budget, stating that (tax) revenues must equal social outlays.

According to (6), those who choose not to work at all receive unemployment benefits reflecting the scarcity value of jobs, and all workers pay for appropriating scarce job assets in proportion to earned income. Someone who works full-time is a net tax recipient or contributor dependent on whether (earning power related) taxes paid (tw) is smaller or greater than the uniform benefits received (pp_c). Now the equilibrium condition for the market for Labour Rights and the balanced budget no longer coincides. Instead, we have two separate conditions:

$$\iint_{e\,w} L\,f(e)f(w)\,d(e)\,d(w) = p \tag{8}$$

and

$$\iint_{e\,w} t\,w\,L\,f(e)f(w)\,d(e)\,d(w) = p\,p_c \tag{9}$$

Equation (8) describes that aggregate labour supply must be equal to the employment level p, and Equation (9) that tax revenues (the LHS) must equal total social expenditures (the RHS). For simplicity, assume that e and w are distributed independently. Substituting (7) into (8) and solving for t gives:

$$t = 1 - \frac{p}{\overline{ew}} \tag{10}$$

and solving (9):[20]

$$p\,p_c = \overline{e}\,t\,(1-t)(\overline{w}^2 + \sigma^2) \tag{11}$$

Substituting Equation (10) into (11) gives the shorthand expression for p_c, presented in the main text:

$$p_c = \frac{(\overline{ew} - p)/\overline{e}}{(\overline{w}^2 + \sigma^2)/\overline{w}^2}$$

For those who want to live merely from the proceeds of selling Labour Rights, the tax rate which maximizes their non-labour income can be derived from:

---

[20] Making use of the fact that $\displaystyle\int_w w^2\,f(w)\,d(w) = \overline{w}^2 + \sigma^2$.

$$\frac{\partial(p\,p_c)}{\partial t} = 0 \Rightarrow t_{\max} = 1/2, \tag{12}$$

corresponding to (see Equation (10)) $p_{\max} = \overline{e}\overline{w}/2$, the same as we found in the case of a homogeneous workforce, except that the uniform wage rate w is now replaced by the average productivity. The maximim objective requires to maximize the income of the least advantaged (the subscript L denotes the lowest productivity level):

$$\max_t \int_e Y_L = \max_t \int_e [(1-t)w_L]L + p\,p_c \tag{13}$$

Substituting (7) for L and (11) for pp_c into (13) and differentiating with respect to t gives:

$$t^* = 1 - \frac{(\overline{w}^2 + \sigma^2)}{2(\overline{w}^2 + \sigma^2 - w_L^2)}, \tag{14}$$

so for $w_L>0$, t^* is lower than t_{max}. It can easily be checked (by solving Equation (14) for $t^* = 0$) that as w_L approaches $\overline{w}$, t^* tends to zero. Similarly, if w_L is set to zero, the tax rate $t_{max} = \frac{1}{2}$ that maximizes the BI is warranted.

A3. The equivalent basic income scheme, uniform productivity

The budget constraint under a BI scheme is:

$$Y = [w(1 - t)]L - B \tag{1'}$$

with B denoting basic income. The labour supply function (2) does not change except that $(w-p_c)$ becomes $w(1-t)$, and instead of Equation (3) we have the balanced budget equation:

$$\int_e t\,w\,L = B, \tag{3'}$$

which gives:

$$B = \overline{e}[t\,(1-t)w^2] \tag{4'}$$

Differentiating Equation (4') with respect to t gives the B-maximizing tax rate $t_{max} = \frac{1}{2}$. Using Equation (2), the participation rate corresponding to this tax rate is $\int_e (L \mid t = \frac{1}{2}) = \bar{e}w/2$, the same value as we found before in the scheme of Labour Rights. The relation between B, L and the tax rate is shown in Figure 2 in the main text.

It can easily be verified that Equation (1') is equivalent with Equation (1) above if tw is equal to p_c and B is equal to pp_c. Denote t_p as the tax rate which realizes a participation rate of p, so:

$$\int_e L \Big| t_p \equiv p \Rightarrow \bar{e}w(1 - t_p) = p \Rightarrow t_p w = \frac{\bar{e}w - p}{\bar{e}} = p_c$$

which proves the equivalence between both schemes.

A4. The equivalent basic income scheme, non-uniform productivity

With unequal talents, we can simply maintain Equations (1') and (3'), but interpret w as a varying parameter. Solving (3') gives:

$$B = \bar{e}t(1 - t)(\overline{w}^2 + \sigma^2), \tag{11'}$$

so the tax rate that maximizes B is $t_{max} = \frac{1}{2}$.

By equating $\int_e \int_w L = p \Rightarrow t = 1 - \frac{p}{\bar{e}w}$, which matches exactly Equation (10) of the Labour Rights system, it follows that labour supply at $t_{max} = \frac{1}{2}$ is equal to $\overline{ew}/2$. Given that both systems are equivalent, it will not come as a surprise that solving Equation (13), with pp_c replaced by B, that is maximizing average income of the group with lowest productivity, yields the same tax rate t^* as expressed in Equation (14).

CHAPTER 4

WHY LAUNCH A BASIC INCOME EXPERIMENT?

1. INTRODUCTION

The limited support for BI and the strong support for workfare-oriented policies (see chapter 1, Table 1) may well explain why there are no BI experiments but many, maybe thousands, welfare-to-work oriented experiments going on. BI and workfare can be seen as opposed ways to achieve more flexible labour markets. Peck and Theodore (2000, 124) assert that welfare-to-work experiments 'seek to articulate a regulatory strategy concerned to make flexible labour markets work. "Work first" approaches, in particular, can be seen as part of a wider attempt to realign welfare provisions, incentive structures and work expectations in light of the "realities" of flexible employment; their aim is to (re)socialise welfare recipients for contingent work.' However, as argued in the previous chapters, a BI can also be seen in the light of flexibility. Under a BI scheme a more flexible labour market may arise because of the elimination of minimum wage legislation and a less comprehensive legislation on employment conditions. By providing a BI unconditionally, the income and utility which potential workers derive from this no-work option serves as a floor. In principle at least, a substantial BI allows the possibility to deregulate the labour market, and in this way to combine the dynamics of American labour markets with the minimum income protection of European welfare states (see also chapter 2).

Comparing BI and workfare (or the shift towards activating labour policies replacing passive welfare), it is interesting to note that a BI experiment may serve as the right counter-experiment for all kinds of workfare-oriented experiments. As argued by Peck and Theodore (*ibid.*, 124-5), the recent popularity of workfarism in the US and the UK can to a large extent be attributed to the positive results of local workfare experiments in the early 1990s. Workfare experiments show the effects of mandatory welfare-to-work programmes compared to the normal treatment (e.g. the duty to apply for jobs, the duty to resume work as soon as possible) of a control group of welfare beneficiaries. Running a workfare and BI experiment simultaneously may show what a difference it makes if recipients must participate, as a condition of income support, in programmes designed to improve their insertion in paid work as under workfare, or if they can freely choose themselves what to do as under BI. It seems reasonable to maintain the services of job training and job counselling even for those receiving a BI if they need help to find a job, although making use of these services is on a voluntary basis. My proposal is to give the experimentals receiving a BI the same per capita value of the cost of these services in the form of a voucher. Because there are no BI experiments going on, we can only guess what the differences would be. For instance, it may well be the case that workfare experiments show better results in terms of labour market inclusion, but that BI experiments show better results in terms of inclusion in all kinds of unpaid work. In any case, comparing the evaluation findings of workfare and BI experiments may give us some information about the effectiveness of welfare-to-work activities performed by employment agencies.

The structure of this chapter is as follows. Section 2 contains a non-exhaustive enumeration of the limitations of existing research to assess the effects of a major change in social security. Section 3 shows the equivalence between a BI and a negative income tax (NIT). Section 4 discusses the New Jersey negative income tax experiments. Although these experiments were held over a quarter of a century ago, some important lessons can still be drawn for new experiments to be initiated in the future. These are presented in section 5. Section 6 presents a structure for a new BI experiment, which can serve as a basis for discussion about any proposal to start such an experiment. In the final section the conclusions are elaborated.

2. THE LIMITATIONS OF THEORETICAL MODELS AND EMPIRICAL RESEARCH

On the basis of *theoretical* microeconomic research[1] something can be said about the direction of the expected effects, but not about the scale of these effects. Although the economic sustainability of a BI is controversial, there are some uncontroversial remarks which can be made. Firstly, the implementation of a BI will reduce the share of GDP which is distributed by the market. Consequently, *on average* a given work effort will be less rewarded when compared to the rewards accruing to the same amount of labour in a scheme of conditional social security. The ultimate effect does however not depend on the higher average tax rate, since the burden (especially of the marginal tax) varies according to one's position in the labour market. Social security recipients now face an effective tax rate equal to 100%, while part- and full-time workers face a much lower rate. The flat tax in the standard BI proposal entails a marginal rate which is comparable across members of society with a low and with a high income. One of the crucial questions therefore is whether the negative effects of a higher average tax rate for the latter group is greater than the positive effects of a lower marginal rate for the former group. Still, even more serious is that economic theory does not yield unambiguous clues about what we can expect for the effect of BI on human capital accumulation (see below), on low wage levels in the absence of minimum wage legislation and on female labour supply.

In the long term at least three effects can be distinguished which will influence human capital formation and hence the distribution of earning powers in the future under a BI scheme. Due to the raising of tax rates required to finance the BI, the net after-tax wage rate will probably be lower under a BI scheme for most workers. This may give a disincentive to invest in human capital, since every unit of human capital will then generate a lower stream of net earnings in the future. However, this is not the whole story. Two counteracting forces work to lower the cost of acquiring human capital: (i) students over the age of 18 engaged in schooling will receive a BI,

---

[1] See e.g. Besley (1990) and Creedy (1996).

whereas most of them now have to incur large debts to finance their study; (ii) with lower net wage rates due to higher tax rates, the foregone earnings of full-time schooling become smaller.[2] Even if we had reliable forecasts about future net wage differentials between educational categories, we would also need to know the effect of monetary incentives on human capital formation (i.e. the allocation of students among educational categories). In sum, one cannot treat earning power, or wage rates, as exogenous in the long run.

It would be helpful if we would know what the effects are of abolishing the minimum wage, eliminating the poverty trap and the effect arising from the absence of preconditions on the behaviour of social security recipients for receiving a social benefit.[3] Not long ago, we have seen a flourishing, yet unresolved, debate on the effect of the level of minimum wages on employment.[4] Note that this debate is about the effect of a small change of minimum wages on employment. What is required here is an estimate of the effect of a complete elimination of minimum wages, in conjunction with the effect of the removal of the poverty trap and making the minimum income guarantee unconditional, on labour demand and labour supply which together will determine the new equilibrium values of wages and employment in the low wage sector under a BI scheme.

Finally, it is much more difficult to model the process leading to changes in the distribution of family income and decisions of family members with regard to labour supply than changes in individual income and labour supply.[5] A standard neo-classical labour supply model where all individuals are taken alike would probably generate entirely different outcomes compared to when one models family behaviour, e.g. when using the male chauvinist model. What is at stake here is the radical uncertainty regarding the effect of a BI on the division of labour within the household. This uncertainty is perhaps responsible for the fact that the feminist movement has not yet taken a clear stance on the BI proposal. There is the fear that the participation rate of women will decline because of the BI: the income loss (or opportunity costs) of not doing paid work becomes less, and a BI can be seen as a disguised wage ('hush money') for housekeeping and childrearing activities.

---

[2] A similar point is made by Atkinson (1995a, 135): 'If the decision is based solely on comparing the expected gain in earnings with the earnings foregone while training, a tax which is simply proportional would reduce both by the same percentage, and the balance in the equation is unaffected. If the tax is at rate t, we would simply have a factor $(1-t)$ appearing on both sides. It is only to the extent that the tax has a graduated marginal rate, falling more heavily on the earnings of trained labour, that the return to training is reduced. Of course this is an over-simplified representation, and costs such as university fees may well not be tax-deductible, but the essential point is that human capital investment largely takes the form of foregone earnings, so that if these earnings would have been taxed, the cost of the investment is reduced as well as the benefits.'

[3] Present social benefits are surrounded by all kinds of obligations (to apply for jobs, to retrain, to fill in forms every month, etc.), whereas a BI can be seen as a kind of (anonymous) gift. The sociological gift-exchange theory predicts that gifts have the tendency to elicit a counter-gift.

[4] See e.g. Kennan (1995), Dolado et al. (1996), Greenaway (1996) and Card and Krueger (1995).

[5] 'To understand family income, one would have to understand not only the process generating other private income sources (dividends, interest and rent) and public income sources, but also the joint decision-making process among family members who adjust their labor supply, human capital, household formation and childbearing decisions in reaction to changes in outside sources of income, as well as to changes in the earnings of other family members' (Gottschalk 1997, 22).

However, even if this fear were realised one can still argue that not much is lost from the perspective of female emancipation. It is likely that a significant part of the decline in women's labour supply would result from women with easy, dead end and low paid part-time jobs quitting their jobs. It is not very likely that women with interesting, well paid full-time jobs would stop working (De Beer 1987, 52-53). Moreover, for both men and women who perform a full-time job it becomes more attractive to work less than full-time: since the share of labour market earnings in household income would become less under a BI scheme compared to the present scheme, the same reduction in working hours will lead to a smaller decline in household income under a BI scheme. This may increase the willingness of men to take a larger share of housekeeping and childrearing responsibilities by working less than full-time. Finally, it may well be the case that because of the abolishment of minimum wages more low wage and part-time jobs will become available for women under a BI scheme. The BI could then be seen as a kind of emancipation fee.[6]

Empirical research on labour supply shows a large variety of outcomes on labour supply decisions resulting from changes in the tax and transfer system. At best, empirical research of this kind can only give reliable estimates for small changes in marginal tax rates, or for small changes in the level of social benefit levels. Attempting to predict the social and economic effects of a major switch from conditional to unconditional social security is a different matter. Therefore, results obtained from empirical research into the effects of benefits, taxes and premiums on labour supply and labour demand only offers insight into the effects of those policy changes which do not cause a fundamental break with the existing system, such as a limited change in benefit levels or tax rates. On the basis of such research no statement can be made regarding the consequences of the completely different arrangement of the social security and tax systems resulting from the introduction of a BI. According to Barry (1997, 161):

> ... no tax and benefit simulation, however conscientiously carried out, can make allowance for the changes in behaviour that would arise under an altered regime. A subsistence-level basic income would face people with an entirely different set of opportunities and incentives from those facing them now. We can speculate about the way in which they might respond, but it would be irresponsible to pretend that by cranking a lot of numbers through a computer we can turn any of that into hard science.

To put it in a terse phrase, there is no hard science concerning the effects of a BI scheme.

Interviewing a representative sample of the population to survey public opinion is likely to be of little use either. We run the risk that people do not answer with complete honesty but give a socially acceptable answer. It is also quite likely that many people do not really know how they would react to the introduction of a BI, because it differs so much from the existing system. It is only when the

---

[6] See Robeyns (2000) for a more extensive gender analysis of the responses of women (specified by groups distinguished by earning capacity and labour market attachment) to the introduction of a basic income.

consequences of a BI are personally experienced that the real meaning of a BI is fully realized and an appropriate answer can be given. The only reasonably trustworthy way to make a statement about the consequences of the introduction of a BI is conducting a field experiment. Such an experiment would involve a limited group of people in a limited area who would, during a limited time, receive a BI. By closely following and analysing the behaviour of this group of experimentals in comparison with a group of controls, not receiving a BI but for instance being subjected to a workfare scheme, we may get some additional insight into the effects of a BI on people's behaviour. Additional, because the information would be complementary to what can be concluded from back of the envelope calculations on the feasibility of BI, and more important, to the findings obtained from sophisticated models simulating an economy with unconditional grants replacing the present scheme of conditional benefits[7] and to the findings of empirical research on labour supply. There are numerous factors at work which influence labour supply decisions. One cannot hope to include all these factors simultaneously within the confines of an economic model. Economic models can, at best, isolate the effects of a few of these factors. An experiment may enable us to solve part of the puzzle, because the limitations of an experiment are of a different nature than those of economic models, whether theoretical or empirical. The main difference is that models rely on assumptions, whereas an experiment allows one to *directly observe* changes in labour market behaviour.

3. BASIC INCOME VERSUS NEGATIVE INCOME TAX

In this and the next section we are mainly dealing with the negative income tax (NIT). Any single tier NIT-scheme can be described by the level of the income guarantee and the withdrawal rate. Both a NIT and a BI provide a guaranteed minimum income to individuals or households, independent of labour market history or current labour market status, and without any work requirement. Whereas the BI is provided to all irrespective of the level of gross income, the level of the NIT depends on gross income. This may seem a large difference, but as Van Parijs (1992, 4) pointed out, both can yield exactly the same distribution of post-tax-and-transfer incomes. This is illustrated in the figure below.

If we take t as the tax rate, y as the gross income, τ as the tax liability, B as the level of BI and N as the guarantee level of the NIT, the tax functions for BI can be written as $\tau_{BI}(y) = ty$ (which corresponds to the horizontally shaded area in panel A of the figure) and for NIT as $\tau_{NIT}(y) = -N + ty$ (which corresponds to the vertically shaded area in panel B of the figure as long as gross income is below break-even and to the horizontally shaded area if gross income is above break-even). Net disposable income y_d can then be written as:

$$y_{d,BI} = y + B - \tau_{BI}(y) = y + B - ty$$

---

[7] See the simulation results of Atkinson and Sutherland (1988) and Atkinson (1995a) using TAXMOD and of Gelauff and Graafland (1994) using the model MIMIC of the Dutch Central Planning Bureau.

and

$$y_{d,NIT} = y - \tau_{NIT}(y) = y + N - ty$$

Figure 1. Basic income vs. negative income tax

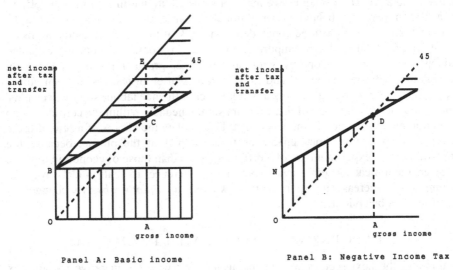

Basic income and negative income tax schemes (Figure 1.1 Parijs (1992, 5), adjusted).

As can be clearly seen, the functions for net disposable income for equivalent (that is, $N = B$ and equal tax rates) NIT and BI schemes are the same. The break-even level of income for the NIT and BI scheme can then be determined by the point at which net and gross income are equal, that is equating net disposable and gross incomes in the functions of net disposable income (graphically, the points of intersection C and D of the bold lines representing the post-tax-and-transfer income at various levels of gross income and the 45 degree line from the origin). For the NIT and the BI scheme the break-even level of gross income equals N/t and B/t respectively.

All single-tier NIT and BI schemes can thus be defined by two variables, the guarantee level and the tax rate. However, under a NIT scheme, the guarantee level is only paid to those individuals or households without any income, whereas the BI received does not depend on the level of gross income. The tax rate of the NIT can be considered as a kind of withdrawal rate as long as gross income is below the break-even level, because the amount of NIT paid by the government is reduced by that rate as income rises. Above the break-even level of gross income, the tax rate is just the normal rate at which gross income is taxed. As can be seen from the figure, the break-even point D is the point at which the individual or household neither receives a transfer payment (the NIT proper), nor has to pay taxes. This corresponds with point C in panel A, where the tax liability CE exactly equals the amount of BI received. The main difference between both schemes is therefore purely administrative, namely whether transfer payments are made *ex ante* (BI) or *ex post* (NIT). In the Introductory chapter, Van Parijs gives three disadvantages of a NIT compared to BI following from the administrative difference. My proposal is to let participants in the experiment choose themselves whether they want a NIT or a BI.

4. THE NEW JERSEY INCOME-MAINTENANCE EXPERIMENT

The New Jersey income-maintenance experiment can be considered as one of the first controlled large-scale field experiments in the field of economics. The details of this experiment are well documented in the three volumes edited by Kershaw and Fair (1976, Vol. I) and Watts and Rees (1977a, Vol. II and 1997b, III).[8] For sake of brevity, I will refer to these three volumes as I, II and III. The principal intention of the experiment was to get reliable information about the work incentives of the non-aged poor under a transfer scheme with an unconditional guaranteed minimum income around the poverty level (I, xiv). At that time, the NIT was taken as a serious alternative to the existing social legislation in the fight against poverty.[9] The prevailing policy to fight poverty used the conventional instruments of education, manpower training and public employment programs. It was a policy based on self-help and self-improvement, and its goal was to make 'tax payers' out of 'tax eaters' (Lenkowski 1986, 39). One of the factors responsible for the rising popularity of the NIT scheme as set out by Nobel laureates Milton Friedman (1962) and James Tobin (1966) was that the spectre of means-testing, causing more harm than good, seemed to come true: '... Negro men were unable to earn enough to support their families and so left home in order that their wives and children might obtain relief' (*ibid.*, 36).[10] The crisis of the welfare state, the welfare 'mess', induced the President to appoint a Commission on Income Maintenance Programs to 'examine any and every plan, however unconventional' (II, xxiii). The sharp rise in the number of people enrolled in welfare programs created the right atmosphere to look for another approach:

> ... rather than making welfare benefits harder to obtain, policy should aim at making it more appealing to give them up. Programs should be designed so that recipients would

---

[8] For other sources, see e.g. Masters and Garfinkle (1977), Burtless and Hausman (1978), Keeley *et al.* (1978) and Munnell (1987). See Widerquist (2004) for a (20 pages long) bibliography, academic as well as non-academic articles, on the NIT experiments.

[9] Two welfare reform bills issued by President Nixon embodying the idea of a NIT, the Family Assistance Plan and H.R. 1, even passed the House in 1970 and 1971 respectively, but both were ultimately rejected by the Senate. For details about the political stratagems going on behind these proposals, two books can be highly recommended: Moynihan (1973) and Lenkowski (1986).

[10] See also the figure in Moynihan (1973, 83) which illustrated the following: 'As male unemployment rates had gone up, so had the number of new AFDC cases. Down, down. Up, up. The correlation was among the strongest known to social science... Then with the onset of the 1960s the relationship weakened abruptly, and by 1963 vanished altogether. Or, rather, reversed itself. For the next five years the nonwhite male unemployment rate declined steadily and the number of AFDC cases rose steadily.'

always have a financial incentive to rely on something other than welfare benefits. Moreover, this should be done not by lowering benefits themselves (which would have been unfair to the 'truly needy'), but preferably by reducing them by less than the full amount of any additional earnings or other income a recipient might have. For every extra dollar or pound gained, the public support a person had been drawing would be reduced by, say, half as much, leaving him better off in total than before. Past a certain point, he would no longer be eligible for any assistance at all and instead would start paying taxes, ideally at the same rate by which his benefits had been lowered. This idea was most widely known as the negative income tax; if it was applied properly, its proponents argued, only those who were really unable to support themselves would remain on relief (Lenkowski 1986, 36).

The failure[11] to reduce (the rise in) the number of welfare recipients, despite the introduction of more severe conditions of entitlement and special programmes designed to help the poor to help themselves, and the rather optimistic view[12] on the behaviour of welfare recipients under a NIT scheme, mobilized enough support for the NIT-experiments.

Before more detailed information about the design of the NIT-experiments in the USA is presented, there are a few peculiarities which must be kept in mind when assessing the relevance of the experiments for the European context. Firstly, the population on welfare at the time the experiment started consisted mainly of female-headed families, since men were not entitled to social assistance (the only ones who could receive benefits were those with an unemployment or disability insurance (I, 9)). This is of course a major difference with the context in which an experiment nowadays would operate. The men enrolled in the New Jersey experiment were confronted with the fact that for the first time in their life they would receive welfare benefits, even if work was voluntarily abandoned, if they had no unemployment insurance and if they were not prepared to do any paid work at all.[13] This may lead to a higher estimate of the negative labour supply response than what we would expect from the introduction of an unconditional scheme today in Europe. Nowadays healthy men passing the means-test without any current income are entitled to social assistance. Those with a high preference for leisure and who have managed to be on the dole cannot reduce their zero labour supply any further when conditional social security is replaced by a BI scheme. The effect of cutting back hours of work among those with a low preference for paid work when easy accessible social security becomes available has to a large extent already manifested itself.

Secondly, the primary focus of the experiment were the labour supply responses to providing unconditional social security to low income *male* earners. Female

---

[11] 'Dependency had become a social condition beyond the apparent power of social policy to affect, save possibly at the margin. *This was the heart of the Administration's understanding of the matter*. It is not a judgement that will be found in the archives. It was not even a judgement. It was simply an awareness of the limits of knowledge that gradually emerged and thereafter did not need to be dwelt upon or even acknowledged' (*ibid*., 353).

[12] The NIT-plan was not 'predicated on the assumption that people don't want to go to work' (Moynihan 1973, 340).

[13] Lenkowski (1986, 56) states that '... one of the bedrocks of the existing policy [was] the tradition of not providing income on the basis of need to those able to work'.

headed families were already entitled to AFDC-benefits (Aid to Families with Dependent Children), and the Social Security Act of 1967 contained a kind of anti-cumulation measure to limit the effective (withdrawal) tax rate on AFDC benefits to 67% (I, 10). Giving these women the opportunity to enrol in a NIT-experiment would probably generate little additional information compared with what was already known. For these reasons, and because the female participation rate was low, it was decided not to include female-headed families in the experiment.

Thirdly, the ethnic composition of whites, blacks and Spanish-speaking of both the treatment and control groups was roughly one-third for each over all cities with a NIT-experiment (I, Table 2.3, 36). Labour supply responses turned out to be significantly different for each of the major ethnic groups (II, 77-85). It is likely that the ethnic heterogeneity of an experiment in Europe will be much lower.

Adding the cultural differences between the USA and Europe, the long time which passed since the experiments started, the gradual improvement of conditions of employment since then,[14] the overall decrease of the working week, and the relatively low poverty level compared with the present social minimum at which the minimum level of benefits is now pitched, means that the outcomes of the experiment are of limited use for answering the question whether the introduction of a BI or NIT around the social minimum in Europe today would have a detrimental effect on labour market participation. Atkinson (1995a, 150) states that 'The NIT experiments are generally considered to have reduced the range of uncertainty surrounding the response of hours of work to taxation...' However, '... there is no necessary reason to expect the results to apply equally in a European context. Those interested in a BI/FT [BI/flat tax] scheme in Europe might like to consider launching such an experimental research project, which would serve both to throw light on the economic effects of the reform and to demonstrate how it would work in reality.'

For readers unacquainted with the outcomes of the experiments, I quote the major findings:

> The most important group... the experiment was specifically designed to examine, is that constituted by the non-aged, able-bodied males with family responsibilities. These are the people with the most labor to withdraw. These are the people about whom there is the most widespread fear that, given an income alternative, they will decide not to work. As it turned out, the effect for this group was almost undetectable... the employment rate for male family heads in the experimental group was only 1.5 percent less than that for the controls. For the number of hours worked per week, the difference amounted to just over 2 percent... The second group in terms of policy interest is the wives. The average family size in the sample was six, so the wives in the experiment were, on average, mother of four children. For this group, the differential between experimentals and controls was substantial, with experimental wives working 23 percent fewer hours per week than the controls, their employment rate being 24 percent less, and their average earnings per week totalling 20.3 percent less. This can be regarded as a desirable outcome, given the fact that wives in six-person families work hard inside the home, and that this work could well be more beneficial (cost-effective)

---

[14] Better conditions of employment reduce the role of net wages as an incentive to elicit work effort and labour supply. Hence, lower net wages due to the higher required tax rates to finance a BI or NIT scheme will have a lower negative effect on labour supply.

from a national point of view than low-wage market labor. It should be noted, in addition, that although this relative reduction is large, it in fact starts from an average figure of only 4.4 hours a week... In the area of psychological and sociological responses, the effects were negligible. Cash assistance at the levels involved in this study does not appear to have a systematic impact on the recipients' health, self-esteem, social integration, or perceived quality of life, among many other variables. Nor does it appear to have an adverse effect on family composition, marital stability, or fertility rates (I, 20-21).

4.1. The design of the New Jersey experiment

Table 1 contains the parameters of the eight NIT-experiments, varying in income guarantee levels and withdrawal rates, which together constituted the graduated work incentives experiment in New Jersey.

Table 1. The Negative-Income-Tax Plans Used in the New Jersey Experiment (I, Table 1.3, 10, adjusted.

	Withdrawal rate (in %)		
Income Guarantee (in % of poverty line)	30	50	70
50	X	X	
75	X	X	X
100		X	X
125		X	

The plans in the upper right corner with high withdrawal rates and low income guarantees are least attractive and in the lower left corner most attractive to participants. To minimize transfer costs (the income support received), low income families were more than proportionally allocated to the least attractive plans in the top rows and higher income families to the most attractive schemes in the bottom rows (I, 13, 96-97). Aside from prospective transfer costs, the assignment of the number of families to each plan was determined by a policy weight given to each plan (the 75-50 and 100-50 plan got the highest weights) and by the expected attrition rate (drop out was expected to be inversely related to the generosity of the plan, and positively related to pre-enrolment annual income).[15] At the end of the

---

[15] The highest attrition or drop out rate, 50%, was expected for the controls, since they would only receive the fees for filling the forms and interviews. This is one of the reasons why the number of controls (632) is almost as large as the number of experimentals (725).

experiment, the overall attrition rate was 20%; 25% for the controls and 16% for the experimentals (I, 105). As could be expected, the attrition rate was higher the lower the guarantee level, the higher the withdrawal rate of the plan and the lower the amount of the last transfer payment (I, Tables 7.3-7.5, 109-111). Most impressive is the amount of effort spent on keeping attrition to a minimum and the quest to recover data on families who left, even outside the USA. At stake here is not only the representativeness of the empirical outcomes of the experiment, but also whether the outcomes could serve as reliable estimates for the costs of a national programme.[16]

Sample eligibility was restricted to families having one healthy man and with a family income below 150% of the poverty line (I, 8).[17] Two reasons were given for this decision:

> First, those close to the field operators wanted to be sure that most of the sample would qualify for significant payments to keep the goodwill of the experimental participants and to minimize the number who dropped out during the experiment. Second, OEO [the official sponsor of the experiment] did not want to be in the position of funding a cash program that was primarily addressed to the nonpoor (I, 10).

Another choice was between running the experiment nation-wide with participants all over the USA or running it in compact geographical areas. There were two reasons that led to the latter choice. First, administrating and monitoring the experiment was much easier when all enrolled families live in the same area. Second, a nation-wide experiment has the disadvantage that participants operate on different geographical labour markets which necessitates to disentangle geographical labour market effects from individual labour supply responses. To compensate for the loss of randomness, New Jersey was chosen because the unemployment rate there was close to the national average.

4.2. The operations, surveys, and administration

The main purpose of conducting an experiment is to collect information. Obviously, this is very costly. The actual (and budgeted) administrative and research costs of the experiment were more than twice the actual transferred income support payments! (I, Table 1.6, 18). The experimentals in the NIT-experiment were interviewed more than twenty times: a (44-question) screening interview to determine eligibility, a (340-question) pre-enrolment before, and a follow-up

---

[16] 'The people who drop out of the experiment may be the same as those who fail to be included in a national program (those who fail to register and those who fail to report their income or in other ways fail to maintain their right to benefits). To the extent that this is the case, the behavioral responses measured in the experiment will be a good measure of the responses that may be expected in the population as a whole. To the extent, however, that the people who drop out of the experiment differ significantly from those that remain and to the extent that they could be expected to be included in a national program, the estimates will be biased in a way that will impair the usefulness of the experiment as a guide to a national program' (I, 117-8).

[17] A major drawback of this decision was that families with both spouses working regularly were underrepresented.

interview after the experiment, twelve regular quarterly interviews, and six special one-shot interviews (I, 15, 24). Those who were interviewed received $5 per interview. Income had to be reported on a monthly basis, and in return for filling in the income report form on time, $10 was paid on top of the normal transfer payment.[18] Before enrolment, the participants received a clear article explaining the working of a NIT-scheme,[19] and an enrolment kit containing the rules of operation, a tax table from which they could read how much one could expect to receive at various income levels, a payments calendar and instructions for filling in the income report form (I, 29). The rationale behind the enrolment kit was to limit as much as possible the contact between staff and families. This was decided in order '... to replicate in so far as possible the operation of a universal negative-income-tax program. One of the important elements of such a system, in contrast to existing public assistance programs, would be this lack of direct contact; it was important to learn the extent to which the families could function without casework contact. In addition, frequent contact could only increase whatever Hawthorne or other experimental effects there might be' (I, 30).

The selection of the sample (both experimentals and controls) was done on a step-by-step basis. From a random sample of nearly fifty thousand housing units only three thousand were found eligible for a pre-enrolment interview. This comprehensive interview further reduced to half the number of those eligible for actual enrolment. In the end, 1357 families, 725 experimentals and 632 controls, were selected (I, Table 2.1 and 2.3, 31, 36). The characteristics of the final sample were compared with the 1970 Census data to ensure representativeness (I, Table 2.2, 34). In addition, a separate check was carried out to compare families who refused enrolment to those who accepted. Only minor differences were found (I, Tables 2.6-2.9, 40-43). A major factor which may explain why some families refused to participate, given their 1968 annual income, was the small initial transfer payment that they would receive for participation.

Concerning the rules of operation of the experiment, three major decisions had to be made: the definition of the family, the definition of income, and the accounting period (I, 75-81). A family was defined as relatives and adopted living with the male family head. Those leaving the family could take with them their part of the guarantee, but could not start a new filing unit. Regarding family income, decisions had to be made on whether or not to include items like gifts and inheritances, dividends, earned interest and rental incomes, life-insurance, rent subsidies, and medical costs. The accounting period was on a four-week or monthly basis.

The questionnaires covered five major (economic and sociological) topics: work and income patterns (about job training, job history, wife's labour force history, child care and welfare history), debts and assets (ranging from property ownership to the net worth of consumer durables), family life and background (about family composition, family planning, educational background, religion, hobbies, etc.),

---

[18] The filing fees were forfeited if the income report form was not returned within four weeks. This filing fee was introduced ten months after the experiment started (I, 54).
[19] J. Tobin, J. Pechman and P. Miezscowski, Is a Negative Income Tax Practical?, *Yale Tax Journal* 77, 1967, 1-27.

political and social life (about political awareness, social networks) and other assorted topics (e.g. social status, self-esteem, worry and happiness, attitudes toward work, and job satisfaction) (I, 149-162).

5. LESSONS DRAWN FROM THE NEW JERSEY EXPERIMENTS

Thanks to the elaborate documentation[20] of the New Jersey experiments, some important lessons can be learned for any new experiment. Firstly, for reasons given previously, the NIT-experiments did not include female-headed families. The reasons for this are no longer valid. Moreover, the female participation rate has risen sharply since then. An experiment today should therefore include female headed families. Secondly, although the USA is a large country, the reasons given above to run the experiment in one confined geographical area seem convincing. Even in smaller European countries there are large differences between labour markets in different parts of the country. Moreover, administrating the experiment nation-wide is probably more costly. Thirdly, the main purpose of the experiment is to collect information about labour supply responses. The quality of this information depends to a large extent on the co-operation of both experimentals and controls for providing timely and accurate information. For this reason, the instrument of filing fees which are forfeited if the required information is not adequate or on time is very useful. Moreover, for the controls and for families with zero transfer payments, the filing fees are the only rewards for participating in the experiment. Rather generous filing fees serve two purposes: (i) to enhance the quality of the information and (ii) to reduce the attrition rate among controls and experimentals with zero transfer payments. Filing fees should of course not be too high, or else they might disturb the behavioural effects of the BI itself. Finally, the researchers of the New Jersey experiments never set out in advance what effects they expected to find.[21] To fill this gap it is important that the main effects to be expected from the introduction of a BI scheme are listed beforehand.

The experiment in the USA was very costly. It is not realistic to expect that a comparable budget will be made available for an experiment in a smaller European country. In order to keep the total cost of the experiment on a modest level the following proposals are suggested. First to take into account that the administrative and research expenses of the New Jersey experiments were more than twice the actual income support transfer payments. To reduce non-transfer costs, especially interview and research costs, we propose not to collect information on psychological and sociological effects since these effects were found negligible. This may almost halve interview and research costs with a likely low loss of useful information. In order to reduce the amount of transfer payments, we propose to include in the

---

[20] Especially the in this paper much cited Vol. I and the appendices, which contain a chronology of events, and official descriptions of the rules of operation for the programme, the definition of income, the definition of the family unit, the filing periods and procedures and the review board.

[21] The researchers involved in the experiment never agreed on, or set down in advance, a summary of what they felt was the most likely outcome for labour supply. 'In retrospect, this is unfortunate. Any attempt to do so now is bound to reflect, to some extent, our present knowledge of the results and thus understate the degree to which we have been surprised' (II, 14).

experiment mainly social assistance recipients and families with an income around the break-even level (see section 6 below). Further, a decision has to be made on the number of plans. The fact that the New Jersey experiment contained eight different plans was mainly due to the fact that at the time it started, able-bodied men were not entitled to social benefits - unless they had social insurance. In these circumstances, and given that the non-poor were excluded, it was not difficult to find enough participants for each plan. In present circumstances, one has to take the prevailing welfare arrangements into account. Now, an individualized BI equal to the social minimum for a single person household would mean that two person households on welfare would gain substantially. This is because the present social assistance benefit for a two persons household in all European countries is much less than twice the benefit for a single person household. At the same time, however, an individualized BI significantly below the present levels of social assistance makes it difficult to find enough participants among single person households. Probably only participants who expect to stop working or want to engage in full-time schooling during the experiment would gain in comparison with what they receive under the current scheme. For these reasons, a differentiated BI, equal to or somewhat below the present social assistance levels, is the most suitable to experiment with. This means that single person and two person households on welfare would receive a BI equal to, or somewhat below, the present social assistance benefit, differentiated for household composition.

Finally, the New Jersey experiments had a duration of three years. An experiment of such a limited duration impairs the reliability of the effects on the labour supply if these are to be translated into a permanent unconditional scheme.[22] Given the budget available for the experiment, the more we can save on administrative, research and transfer costs, the longer the duration of the experiment can be, and the higher its reliability for assessing the effects of a permanent BI scheme.

6. DESIGN OF A NEW BASIC INCOME EXPERIMENT[23]

The purport of this section is to outline the structure of a BI experiment, taking into account the insights obtained from earlier experiments, and to arrive at a better proposal by offering the opportunity to criticize our proposal. As a result, we hope that at the time some country or city is prepared to conduct such an experiment a well constructed plan will be available.

---

[22] 'Consider first the male household head with a steady job involving hard work and long hours. If he knew that negative tax payments were permanent, he might instead take a job with lighter work and more normal hours. Yet for a period of three years, such a shift might seem too risky. At the end of the experiment, he would need the higher earned income but might be unable to get his old job back. For the steadily employed male head, the probability is that an experiment of limited duration will have smaller effects on labor supply than will a permanent program... Wives, teenagers, and other adults in the household are likely to be in and out of the labor force as family circumstances change. To the extent that periods of withdrawal from the labor force are planned in advance, a temporary experiment encourages the concentration of such periods during the experimental years, when the costs of not working are lower than normal' (II, 10).

[23] This section is written in close collaboration with Paul de Beer.

There are three reasons why a field experiment cannot replicate the real introduction of a universal BI:

1. In a limited experiment it is not possible to change the external circumstances in the same way as would be the case with the real implementation of a BI. These circumstances include changes in minimum wage, the wage cost structure of employers, the labour demand (employment levels), shifts in economic structure between different branches of industry, shortening of labour time etc. Since the behaviour of (potential) workers (those working and those seeking work) is to a high degree determined by restrictions from the side of labour demand (i.e. the availability of jobs), the behavioural reactions to a BI will only manifest themselves to a degree in a limited experiment. On theoretical grounds it may be expected that a BI will result in a decrease in labour supply when calculated in average working hours per worker, but an increase of the labour supply when calculated in the number of employed persons. This means that the available work will be spread over a greater number of people. Since the first effect does not occur for the economy as a whole, participants (especially those who otherwise would be involuntary unemployed) in the experiment do not 'profit' from the dilution in the labour supply measured in hours per worker, which would occur with a universal implementation of the BI.

Another effect to be expected, which would not occur in a limited experiment, is that the net income (net wages plus BI) of workers with relatively unattractive work will increase in comparison with the net income of workers with relatively more attractive work. If everyone receives a BI, employers would need to attract employees for unattractive work by means of a higher wage or improved labour conditions. Since only a small number of people participate in the experiment, employers can attract unemployed people who do not participate in the experiment. As these are obliged to accept suitable work or else be penalised, employers would not have to increase wages or improve labour conditions.

2. In practice an introduction of a BI will mean an income improvement for some groups and an income reduction for others. To clarify this point, both systems are graphically illustrated in Figures 2 and 3 (for a single breadwinner family who enjoys twice the tax allowance V) with the gross income on the horizontal and net income on the vertical axis. The line OJAC shows the possible combinations of gross and net (labour) income for workers and people who have no right to a benefit (mainly dependent housewives), SAC shows the gross-net trajectory of beneficiaries, both under the present system which is characterised by a tax free base of OV and a benefit level of OS. For example, someone with a working partner and a small part-time job earns a gross income of OV. Due to the tax-free base this person is not required to pay any tax, but he or she will have to pay tax as soon as the gross income is higher than OV. Figure 2 also illustrates one of the advantages of a BI over the present system, namely the elimination of the poverty trap SA. Under the present arrangement of social security the effective (marginal) tax rate on social assistance recipients is very high. This is partially due to the fact that supplementary arrangements such as rent subsidies and child care subsidies are dependent on income, and also, because working incurs additional costs. An

effective withdrawal rate of about 100% or even higher, is not exceptional. As long as the gross labour income of someone with a welfare benefit is lower than OE not much will change in net terms because the benefit and subsidies are deducted pro rata to the earnings. Finally, under the system of BI every adult receives an amount of OS or OB_s, with B_s the individual BI, but the tax rate on labour income will be higher than the average rate under the current system. Therefore the line B_sD is flatter than the line JAC.

Figure 2. Conditional social security (SAC or OJAC) and unconditional social security (SD).

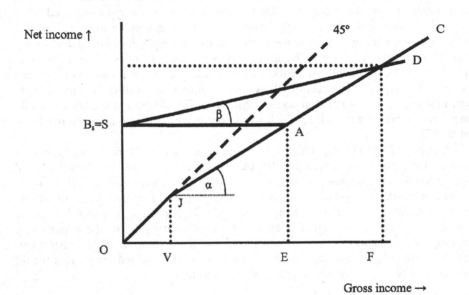

The net income of breadwinners with a gross income of OF is not affected (see Figure 3), no matter which system is adopted (of course, the level of OF depends on the tax rates under both systems). Those with a gross income lower than OF will improve their net income under a BI scheme. For the experiment this means that it is unlikely that people whose income deteriorates because of the introduction of a BI (those with a gross income higher than OF) will be willing to participate in the experiment. Those who are prepared to do so, for instance because it enables them to work less with a small sacrifice of income, do not form a representative sample of the group that would suffer from the introduction of a BI. Therefore, only those for whom the introduction of a BI has a neutral or positive effect will be included in the experiment. It is no problem that especially in this group undesirable behavioural effects are to be expected. People whose situation improves due to a BI would, in the opinion of many opponents of BI, withdraw or partially withdraw from the labour market. In any event the experiment will show whether this effect does indeed take place. However, the experiment could become rather costly especially if people who profit financially from a BI take part in it. If this does not form an insurmountable budgetary problem, then there is no objection to letting these people (all with gross incomes between O and F) take part in the experiment. If the available budget is limited it would be preferable to let only those participate who hardly profit or suffer from a BI (that is, those with pre-enrolment gross income near OF).

Figure 3. Conditional social security (OJC) and unconditional social security (SD) for a breadwinner family.

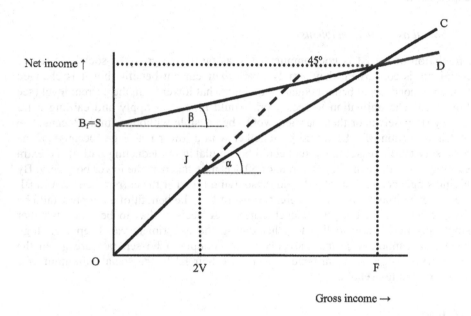

3. Changes in the tax, premium and subsidy schemes that go hand in hand with the introduction of a BI, are difficult to simulate. In a limited experiment it will not be possible to change tax rates, premium levies, and the issue of subsidies. Therefore their effect will need to be represented in the level of the BI itself. As a consequence of this, for many participants in the experiment the BI will not have the characteristic of a constant and unconditional amount but of a benefit or subsidy which is greatly dependent on personal situation and individual behaviour. Taking the previous point into consideration, this means that in practice the BI will have a largely fictitious character for a large group of people. The simulated amount of the BI is cancelled out by the simulated increase in taxes or the reduction in subsidies or benefits. To be sure, even in the existing situation different payments and deductions on the payslip have for most people a fictitious character. This will apply even more with the BI experiment, because officially there are no changes in the tax deductions. By giving the participant regularly a summary of their simulated income situation (e.g. a monthly summary with their fictitious BI, (extra) taxes, subsidies, wage, etc.) the BI can be brought as close as possible to reality.

On the basis of all these considerations, it would seem advisable to limit the experiment to those groups of the population for whom a BI can be relatively simply simulated and without much costs and from whom substantial and – taking the BI discussion into consideration – relevant behavioural reactions can be expected. The two groups that meet these criteria are beneficiaries of minimum level benefits (especially social assistance recipients) and workers who would neither see improvement nor deterioration in their incomes by the introduction of a BI. In addition prospective entrepreneurs could be considered for inclusion in the experiment.

6.1. Social assistance recipients

Simulation of the BI is not complicated in as far as the group of social assistance recipients is concerned: they simply retain their current benefits but it is changed into an unconditional benefit equal to, or somewhat lower than, the current level (see Figure 2). The obligation to seek work would no longer apply and earning extra money (themselves or their partner) would be possible without a limit. Because the real introduction of a BI would go together with a lower net wage (because of the necessary tax increase) this effect could be simulated by deducting part of the extra earnings from the benefit. A tax rate of 50% can be taken as the lower boundary. By giving single person households and those living together the current benefit as a BI, we can experiment with two different levels of BI. The benefit of a couple would be divided into half a benefit for each partner. A decision has to be made whether single people starting to live together during the experiment can keep their high benefit or cannot. It seems unavoidable that couples who separate are given the opportunity to get an extra benefit to top-up their BI to the social minimum of a single-person household.

6.2. Workers

For workers at the break-even level initially nothing changes when they receive a BI. This break-even level is defined as the gross income at which it makes no difference in terms of post-tax-and-transfer income whether one is subject to the present conditional scheme or to an unconditional scheme.[24] Only workers whose income is around the break-even level can participate in the BI experiment. The BI that they receive is compensated by the extra tax they would have to pay with the introduction of a BI (among other reasons because the tax free amount and the transfer of the tax allowance among partners would cease and because of loss of earmarked, but income-dependent subsidies). They would only notice the existence of the BI if something in their situation changed, e.g. if they began to work less. Therefore, at the start of the experiment, they would receive an explanation of the

---

[24] To determine the break-even level, define t' as the tax rate and V as the tax allowance operative under the conditional scheme, and t as the tax rate under a scheme with a BI at level B. The break-even level is then determined by the equation which equates net income at the different schemes, holding gross income (y) constant: $(1 - t')(y - V) + V = y + B - ty$, so $y = (B - t'V)/(t - t')$.

consequences for their net income resulting from different decisions that they could take during the experiment. If for example someone with a full-time job worked one day less per week, this would lead to a less than proportionate decrease in net income, instead of an approximately proportionate income decrease of about 20% which would be the case under the current system. So the experiment acts as a stimulus for participants to work less.

Somewhat problematic is that those who start to work longer or receive a promotion will earn less in a BI system than in the current system. Because of the previously mentioned reasons, it is unlikely that this effect can be expressed in the experiment. For this person, with a gross income now above OF, it would have been better not to participate in the experiment. Therefore a simulation of the effect of a BI will not be possible when an income improvement occurs. On theoretical grounds it is unlikely that in the short run the introduction of an income-neutral BI will lead to a desire to work longer (the so called income effect is zero while the substitution effect is negative) so the lack of this (possible) aspect does not seem too serious.

Another problem presents itself with workers who are sole providers. With the real introduction of the BI the non-earning partner would receive her own BI which would be coupled with an income reduction for the breadwinner (who, it is true, would also receive a BI but who would have to pay more tax because of the cessation of breadwinner allowances). Even if total net family income remains the same, it is probably not possible to simulate the income transfer from the breadwinner to the dependent partner. The breadwinner would have to be contractually obliged to transfer part of his wage to his partner: it is not likely that many traditional breadwinners would be prepared to do this (and if they do, they would not form a representative sample). In such cases the BI of the partner would have to be fictitious and only play a role if the partner started working, in which case we face again the problem that if the breadwinner does not work less time than before, the family income would, as a result of the BI, increase by less than under the present system.

The BI can be well simulated with the so called one and a half earners, that is with couples of which the man has a full-time job and the woman a part-time job (or the other way around). It is often expected that these women will cease working, because even without work they would receive their own income. If the introduction of a BI works neutrally for two-earner families, their situation will improve compared with the current system if the woman stops working. If this behaviour manifests itself during the experiment on a large scale (as opponents of the BI expect), it will become rather costly because all women who stop working receive a BI without consequences for their partners income.

6.3. Prospective entrepreneurs

Beside the two categories mentioned above (beneficiaries and workers or families for whom a BI has a neutral effect) a third group, prospective entrepreneurs, could be considered for inclusion in the experiment. According to its proponents a BI stimulates prospective entrepreneurs because initially it is not necessary to earn a full wage. If revenues are equal to costs and no profit made, they still have a BI to live from. The population could be called upon to apply to participate in the experiment if they are interested in starting their own business. Their present social economic position (employed worker, beneficiary, or housewife) would not be of concern (as long as one is not already independent). Not all of them receive a BI, because participants have to be coupled with other applicants who would serve as a control group.

The criteria that must be met have to be clearly defined to prevent the experiment from being misused. For example, workers would be required to terminate their employment (to prevent them from receiving a double income) and housewives will need to prove that they have started their own business (to prevent them from using the BI as a housewife wage). The BI of those who close their businesses during the experiment will have to be stopped (to prevent sham businesses). Furthermore the BI for new independents will be lowered if they make a profit in order to simulate the effect of higher tax rates.

6.4. The cost of the experiment

The suggested limitation of the experiment to only three groups, namely social assistance beneficiaries, workers with a break-even gross income, and prospective entrepreneurs, is motivated by the wish to keep the cost of the experiment as low as possible (or, with a given budget, to make the number of participants as large as possible and to extend its duration), and by research-technical reasons.

In the experiment, the benefit of beneficiaries is replaced by the same amount (or a somewhat lower amount) as a BI so that initially there will be no extra costs involved. The government may well lose revenue because the occasional, extra earnings of participants in the experiment can no longer be deducted from their benefits. However, these losses do not add to the expenditure cost of the experiment. If the BI proves to be a stimulus to work legally instead of in the black economy, as may often be the case at this time, the government's revenue may even improve.

Initially there are no extra costs for workers with a gross income at the break-even level. They do not experience a change in their net income if they do not change working hours. Their BI is financed from the higher tax rates they pay when they participate in the experiment. Extra costs only occur when participants in this group decide to work less. The maximum size of these costs is equal to the number of participants multiplied by their BI. This cost would only be incurred if all the participants in this group decided to stop working. For the group of prospective entrepreneurs, the costs which burden the budget of the experiment apply right from the start. Workers previously in permanent employment or people without an income who participate as entrepreneur in the experiment, receive a BI every month. Any extra deduction on their profit will at best only partially compensate for this.

6.5. Effects of a basic income to be researched

The emphasis in the experiment will be on the labour market effects. There is much disagreement in this area, and these effects are of great importance for determining the feasibility and desirability of a BI. In concrete terms the research is about the consequences of BI on labour supply. Will workers work less or will some people even stop to work? Will beneficiaries be more prepared to accept low paid or part-time jobs? Will non-participating partners ('housewives') seek and take more or less paid work? Will a BI cause interesting behavioural reactions in all sorts of other areas? Although the latter may not be crucial for the judgement of BI's feasibility, it is nonetheless worth watching. For instance, the effects of a BI on consumption patterns, family composition (living together or separate), role division between men and women, the way leisure time is spent and things like participating in volunteers work, social participation, etc. are all possible consequences worth taking into account.

Admittedly, the information that a limited field experiment such as the one suggested here can furnish falls short of what we need to know to say something relevant about the feasibility of the BI proposal (see section 2 above). Still, depending on the outcomes, the experiment may provide some useful information. If those receiving a BI reduce their labour supply significantly in comparison with the control group, then the opponents of BI have, to say the least, the benefit of doubt in their favour. If experimentals and controls do not show any significant differences in behaviour, then it cannot be concluded that a BI has no adverse impacts on total labour supply. It may indeed well be the case that the experimentals perceive the duration of the experiment too short to change their behaviour. Finally, if experimentals, especially those for whom the conditional social assistance benefit is replaced by a BI, show a marked increase in labour supply in comparison with the control group, then this suggests that the bite of the poverty trap is serious and this would strengthen the plea for a BI. The experiment may show that the opponents of BI are right, but it cannot conclusively decide the same in favour of the advocates of BI.

SUMMARY AND CONCLUSIONS

If the idea of a BI is ever going to come high on the political agenda, we have to know which kinds of social and economic effects can be expected by its implementation. Some economic models try to address this issue, but outcomes are very sensitive to how the labour market is modelled and what model makers believe to motivate people. A limited field experiment may enable us to fill part of this gap, because the limitations of empirical research and economic models is of an entirely different nature than the limitations of a real life experiment. For instance, in the former type of analyses, it is implicitly assumed that parameters describing the labour market behaviour of economic agents remain constant. Moreover, these estimated or calibrated parameters are obtained by imposing a particular labour supply function on agents. The advantage of a real life experiment in this respect is that we can directly measure the labour supply responses of experimentals compared to those in the control group. Contrary to the results obtained from empirical research on the labour supply responses of some group to non-labour income, we do not need to extrapolate the results to other groups, nor are the outcomes influenced by what model makers believe motivates people.

The study of the BI proposal is therefore surrounded by great uncertainties with respect to the changes in citizens' patterns of behaviour in response to such a reform. I doubt whether any firm conclusions can be drawn from either theoretical models or empirical research which try to scrutinize the effects of a substantial BI. With this as the point of departure, three interrelated issues were addressed. Firstly, I argued that a limited field experiment of a BI may solve part of the puzzle concerning economic feasibility. Secondly, any new experiment should be held informed of the New Jersey NIT experiments. Although the outcome of these experiments cannot be considered representative for the expected effects of the introduction of a BI in Europe or even the USA today, some important lessons can be drawn for setting up a new experiment. Finally, field experiments are rather costly ways to collect information, and thereby in designing this new experiment, we have to sail between Scylla and Charybdis. We must not allow experimental costs rise too high, and we must try to collect as much relevant information as possible. Hopefully the proposed design of an experiment may help to overcome some obstacles to the launching of a BI experiment somewhere in Europe, alongside the many workfare-oriented experiments already in place.

CHAPTER 5

FIRST STEPS TOWARDS A BASIC INCOME [1]

1. INTRODUCTION

In this chapter the first steps are presented that transforms conditional social security in a gradual way into unconditional social security. The problem social policy makers face in fighting unemployment is that the social minimum must be kept at a reasonable level, large scale unemployment (especially at the lower end of the labour market) must be reduced, under the restriction that whatever policy measures are taken, there must remain sufficient (overall) incentives to work. Advocates of a BI assert that if we want to maintain the social minimum at a reasonable level, and at the same time want to reduce large scale unemployment at the lower end of the labour market (through gradually abolishing minimum wages and strenthening monetary incentives to work by eliminating the poverty trap), we inevitably move towards a BI-type scheme. I will therefore concentrate on what are probably the two most important adverse side-effects of the present welfare state: firstly, the poverty trap and secondly, unemployment at the bottom end of the labour market due to binding minimum wages. Removing both may be valuable in itself, but also brings the present system closer to that of a BI system.

The poverty trap is to a large extent due to the conditional nature of the present social security system.[2] Recipients face a high withdrawal rate on all net labour income up to the social security benefit, working may engender additional costs (e.g. travelling expenses, costs of day-care centres, etc.) and housing rent subsidies or subsidies for the use of day-care centres may be reduced when income goes up. In practice, an effective marginal tax rate of 100% or more is no exception. Those who have few marketable assets are being trapped in poverty. Moreover, the opportunities for them to find employment are seriously damaged by high minimum wages. Those captured within the poverty trap experience a strong disincentive to work: they can only escape the poverty trap when they succeed in finding a job with gross-of-tax earnings considerably higher than the level of their benefits. All (part-time) work with gross-of-tax earnings below a certain threshold will not yield any, or only very modest, monetary rewards for those who also receive a means-tested benefit. This may partly explain why part-time work is mainly undertaken by women. Most women with a working partner are not entitled to any benefit. Therefore they do not face the poverty trap and subsequently any part-time work will raise family income. As we will see in sections 3 and 4, the implementation of a flat-tax BI will eliminate this poverty trap.

---

[1] A previous version of this chapter was published in *De Economist*, 1997, 145 (2), 203-227 under the title 'An Alternative Route to a Basic Income: The Transition from Conditional to Unconditional Social Security'.
[2] See Whynes (1993).

115

A second problem that most modern welfare states have to deal with is unemployment. Although unemployment is prevalent among all categories of workers ranging from the lower to the higher educated, its incidence is particularly strong among those with low earning power, a low educational status and the least fortunate background characteristics. Traditional text book economics sees unemployment as being partly due to the relatively high minimum wages which exist. A minimum waged marginal worker is that person who has a labour productivity which is equal to the gross-of-tax minimum wage. All workers must have a net labour productivity greater than or equal to their minimum wage. All potential work with a wage rate lying between zero and the minimum wage will either not be undertaken or will be carried out within the moonlighting sector. What are the employment perspectives for the long-term unemployed and for those with the least productive skills under a BI? In theory, a BI regime will offer employment opportunities for all those with non-negative labour productivity. In short, the argument is as follows. Under a BI regime there is no need for minimum wages, because all (potential) workers already receive a BI as a fall-back option. Therefore, under a BI system it is profitable for both employers and workers if those people with positive productivity levels do work. Under a BI scheme the marginal worker is any person who has a positive productivity level, rather then a productivity level equal to the gross-of-tax minimum wage.

In the remainder of this chapter I will propose some measures which eliminate the poverty traps and the bite of the minimum wage on employment of the lower skilled. At the same time these measures can be seen as first steps towards a full-fledged BI system. In section 2 some remarks are made with respect to the so called impossibility theorem: A BI is either too low to be socially acceptable or too high to be economically feasible. Section 3 looks at the royal way towards BI, namely when the present system of social security is transformed into a BI system by means of a gradually increasing partial BI. Section 4 provides an alternative route, section 5 deals with the position of part-time workers and section 6 discusses a differential BI dependent on household composition. Conclusions can be found in the final section.

2. THE IMPOSSIBILITY THEOREM: A BASIC INCOME IS EITHER TOO LOW TO BE SOCIALLY ACCEPTABLE OR TOO HIGH TO BE ECONOMICALLY FEASIBLE

The above title verbalises a wide-spread *belief* that a BI is rather utopian. One may agree with some of the favourable aspects a BI has to offer but still be opposed to its implementation due to its impracticability or insustainability. Opponents of a BI assert that if the BI is set at the present social minimum level, the total reduction of labour supply will be so large that it will result in a reduction in GDP. Total tax revenues needed to finance a BI would be so high (e.g. 25-30% of GDP) that the level of the flat income tax rate would also have to be high (more than 50%). There is also the danger of a downward spiral: if overall labour supply is reduced, because of a reduction in labour supply among the low and middle income earners (due to a reduction in their after-tax wage rate) which is not compensated by an increase in labour supply among the unemployed who now face better incentives, GDP falls and the flat tax rate would have to be even higher to raise sufficient revenues for the BI fund. To avoid this danger, a BI would have to be set at a level which would not reduce GDP. Taking this into account, opponents believe that a BI compatible with the present level of GDP is too low to be socially acceptable as a social minimum. In short, opponents conclude that a BI would be either too low to be socially acceptable, or too high to be economically feasible.

The economic sustainability of a BI is a controversial matter. Whatever can sensibly be said about the economic feasibility of a full BI at the present state of knowledge, it is for certain that its implementation is not an overnight affair.[3] However, certain trends within society may make it easier to introduce a BI, or may make its introduction more urgent. Since World War II most western countries have experienced continuous growth in GDP per capita, a decrease in the average number of hours worked, a gradual improvement of the conditions of employment, an increase in the female participation rate and an increase in the number of unemployed, disabled and early retired people entitled to a social security benefit. All these factors facilitate the implementation of a BI. High unemployment exerts pressure on working hours, reducing them so as to distribute the total amount of employment over more people, and in effect this can be used to accommodate the introduction of a BI. The gradual improvement of the conditions of employment, combined with the shortening of the working week, makes it all the more feasible to reduce labour income as a necessary compensation for the sacrifice endured. What makes the introduction of a BI so expensive is the large number of women who do not perform paid work and are not entitled to a social security benefit under the present arrangements. If the female participation rate continues to rise in the future, it will become less expensive to finance a BI, provided a BI will not deter women from the labour market. At the same time, the larger the fraction of the work force receiving a conditional social security benefit, the more the economy suffers from the adverse effects of the poverty trap. Not unimportant, a BI might reverse the trend of rising income inequality in the last two decades. Finally, the ratio of the social minimum to the average net income is not fixed, due to the fact that a large part of the social minimum is a biological minimum (e.g. food, shelter). As GDP per capita increases, a greater proportion of it will be spent on luxuries (those goods which have an income elasticity greater than one, e.g. winter sport holidays or mobile phones). The purchasing power of a BI need not include these luxuries.

The following sections compare two alternative ways of implementing a BI. The normal route, outlined in the next section, is to start with a partial BI which is gradu-ally increased during a long period of time. This approach has several disadvantages: social security becomes a mixture of two regimes, incentives to perform paid work for persons neither performing paid work nor entitled to social security benefits are reduced, the partial BI paid to this group is to a large extent unnecessary from the perspective of guaranteeing a minimum income, and finally it

---

[3] In the following sections I argue for a very gradual route of implementation where the basic features of conditional social security are maintained during the first two phases.

does not reduce the poverty trap from below, but from above. The alternative route, expound in sections 4 and 5, uses three characteristics of the present conditional social security system (the tax allowance, the minimum wage and the withdrawal rate) to implement gradually a gross-net earnings trajectory which closely resembles the gross-net earnings trajectory belonging to a BI system.

3. A PARTIAL BASIC INCOME

Suppose that the government and the political parties which are in power are convinced that a transformation from conditional to unconditional social security is desirable in the very long term, acknowledging the adverse effects that the present conditional social security system has on the economy, e.g. the poverty trap, relatively high minimum wages (where no labour is demanded when the hourly wage rate lies between zero and the minimum wage), the dangers of a downward spiral[4] and a split society when too many able-bodied citizens wish to work but cannot do so etc. Altogether they believe that in the long term a transition towards unconditional social security might be desirable.

The problem is the transition from conditional to unconditional social security in the long term. The immediate implementation of a full BI scheme would give the economy too large a shock and would cause unacceptable and sudden changes in the distribution of net disposable incomes. The present social security system needs to be transformed to a BI regime with as few (unexpected) shocks and (unintended) redistributional effects as possible. According to the BI literature the correct way to make this transition is to start by implementing a *partial* BI and then to increase it incrementally until the full BI level is reached (defined as a fixed percentage of average income per capita, say 25%).[5] The gradual increase of the partial BI will be accompanied by a gradual increase of the tax rate along with a gradual decrease of the minimum wage. During this transition the tax rate is always set at a level high enough to finance both the growing expenditures needed for the disbursement of the partial BI and the necessary supplementary conditional benefits required to keep those with very low incomes or no income at all at the prevalent social minimum level.

The advantage of this approach is that the government during transition can check whether the measures have positive effects. If the expectations were too optimistic, the government can eventually decide to aim at a partial BI which falls short of the prevalent social minimum level. The disadvantage of this approach is that social security becomes a mixture of two regimes: conditional and

---

[4] When unemployment and disability rates are high, together with high unemployment benefits and disability payments, gross-of-tax labour costs (net wages plus taxes including social security contributions) will also be high. For the employer it is rational to hire more workers as long as their marginal productivity is at least as high as the gross-of-tax labour costs. As more people become unemployed or disabled, labour costs will rise due to the higher taxes which are needed to finance the higher social security outlays. But higher labour costs will lead to more lay-offs. The productivity of marginal workers will no longer be high enough to outweigh the higher labour costs. This results in a downward spiral during economic downturns and an upward spiral during economic upturns.

[5] See e.g. Heij *et al.* (1993), WRR (1985) and Centraal Planbureau (1992).

unconditional. As long as the transition is not complete, the government will keep a large administrative apparatus to control citizens' claims on the supplementary conditional benefits. But besides the forgone reduction in administrative costs this approach suffers from other disadvantages. The second disadvantage is that a partial BI is given to all adult citizens, whether working or not, whether rich or poor, thus including those who have enough means (family income) to provide themselves an adequate level of subsistence. One major factor which makes the implementation of both a partial and a full BI so expensive is the large number of citizens who are at present neither performing paid work nor entitled to a conditional social security benefit. This group is mainly populated by dependent housewives. As stated above, this group does not face the poverty trap. One may even expect a (partial) BI to reduce the incentives for this group to perform paid work, as on balance the negative income effect of an unconditional BI dominates the substitution effect with respect to their labour supply.[6] A BI paid to this group is to a large extent unnecessary as family income will be more than sufficient in most cases. Thus the transition from conditional to unconditional social security through the use of a partial BI scheme may lead to an immediate infringement of one of the basic principles of the present social security system (the principle of selectivity). Moreover, present social security benefits are only meant to be temporary while the BI is a permanent benefit. Thirdly, a partial BI does not reduce the poverty trap from below, but only from above. To observe this consider Charts 1 and 2.

In Chart 1 conditional and unconditional social security systems are represented in their pure forms (in all schemes the social minimum is taken as given).[7] The horizontal axis measures gross-of tax labour income and the vertical axis net-of-tax and transfer income. The line SD represents the gross-net traject of a pure BI regime with a single proportional tax rate on all income (equal to 60% here, whereas the balancing budget tax rate for the first system is assumed to be 25%). The line SAC represents conditional social security, which is characterized by a gross and net minimum wage (y_m and M, respectively) and a withdrawal rate of 100%. The unemployed receive a social security benefit equal to OS, and additional earnings up to a certain level will be retained on their benefit. With a tax allowance equal to V and a single proportional tax rate equal to t, gross-of-tax earnings y_p (which correspond to point A) which equates net-of-tax labour income to the social minimum can be derived from the formula:

$$(1-t)(y_p - V) - V = S = \quad y_p = (S - tV)/(1-t) \qquad (1)$$

---

[6] With a BI, the income effect on labour supply is negative for those with low labour income or no labour income at all. The sign of the substitution effect is more difficult to predict for this group. Women with a working partner face the same marginal tax rate as their partners as tax liabilities depend on family income. But the implementation of a BI will also raise the (marginal) tax rate. If the marginal tax rate for women is more or less equal in both regimes, the substitution effect will be close to zero.

[7] Highly stylized graphs are used to characterize both systems. Admittedly, welfare arrangements in the real world are much more complicated, but as long as welfare states incorporate minimum wages and maintain a social minimum the basic idea of an alternative route towards a BI is valid.

I assume the social minimum is a fixed percentage f of the net minimum wage M:

$$S = fM. \tag{2}$$

Given M, V and t it is easy to derive the gross minimum wage y_m:

$$(1 - t)(y_m - V) + V = M = y_m = (M - tV)/(1 - t) \tag{2a}$$

Now that we have derived the poverty trap and the gross minimum wage we can locate the different categories in the labour market:

1: persons without earnings and not entitled to social security benefits in O;
2a: social security recipients with gross-earnings y equal to zero in S;
2b: social security recipients with positive gross-earnings $y < y_p$ on SA;
3a: minimum wage earners in H and workers with gross earnings greater than y_m on HC;
3b: part-time workers on OJAC or SAHC.

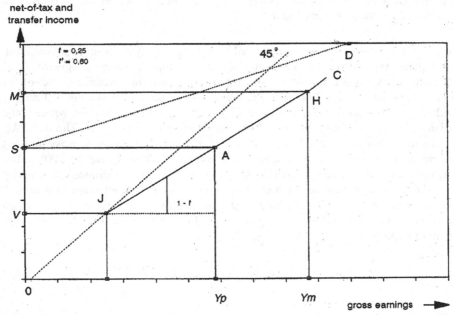

Chart 1 – Conditional social security (SAHC) with poverty trap (SA) and minimum wage and unconditional social security (SD)

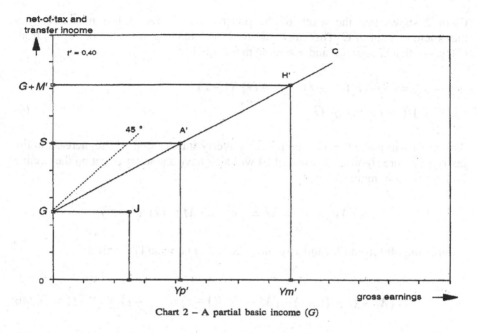

Chart 2 – A partial basic income (G)

As a net minimum wage M exists, all full-time workers will have gross labour incomes greater than or equal to y_m. Between O and this point on the horizontal axis only part-time workers exist. They work at least against the hourly minimum wage rate but their total labour income will probably not often exceed y_m. To escape the poverty trap the unemployed must find a (part-time) job with gross earnings greater than or equal to y_p. All employment with earnings below this point are only financially attractive for those not receiving means-tested benefits (e.g. dependent housewives) and therefore not subject to the poverty trap.

Now consider what happens if a partial BI G is substituted for the tax allowance (Chart 2). All adults will receive a BI equal to G, but the unemployed will receive supplementary conditional benefits equal to (S-G). Group 2a will stay at point S, but group 1 will move from point O to point G along the vertical axis. To finance this partial BI the tax rate must be raised from t to t'.[8] The width of the poverty trap (SA') can be derived from the formula which again equates net-of-tax labour income to the social minimum:

$$(1-t')y_p' + G = S \Rightarrow y_p' = (S-G)/(1-t').$$ (3)

---

[8] For illustrative purposes it is assumed that for all variants the government works with a balanced budget. Whether the budget balances or not cannot be seen from the figures, as they do not show how many people are located at the different points.

Chart 2 shows that the width of the poverty trap is reduced from above by a maximum amount of G. This can be derived by deducting the outcomes of (3) from (1), given that G equals V and assuming that t equals t':

$$y_p - y'_p = (S - tV)/(1-t) - (S-G)/(1-t) =$$
$$(G - tV)/(1-t) = G \text{ if } G = V. \tag{4}$$

Any $t' > t$ will make the reduction in the poverty trap smaller. If the increase in the tax rate is more significant, a partial BI will also have a smaller effect on the decline in the gross minimum wage:

$$(1-t')y'_m + G = M \Rightarrow y'_m = (M - G)/(1-t'). \tag{4a}$$

Subtracting (4a) from (2a) and assuming again that $G=V$ and $t=t'$ gives:

$$y_m - y'_m = (M - tV)/(1-t) - (M-G)/(1-t) = (G-tV)/(1-t) = G. \tag{4b}$$

Due to the fact that a partial BI reduces the poverty trap from above, the trap (SA') remains for the gross earnings section of O to y_p'. Transforming the present conditional social security scheme to an unconditional social security scheme by means of a partial BI has four main disadvantages: social security becomes a mixture of two regimes, incentives to perform paid work for persons belonging to group 1 are reduced, while at the same time the partial BI paid to this group is to a large extent unnecessary from the perspective of guaranteeing a minimum income and finally a partial BI does not reduce the poverty trap from below, but from above. The main problem is that the partial BI scenario is in conflict with the basic principles upon which the conditional social security is built. The substitution of the tax allowance for the partial BI means that some fundamental basic features of the BI proposal are directly imposed on the social security system at the expense of some fundamental basic features of conditional social security. However, an alternative scheme of transition towards unconditional social security which departs more from the present social security arrangements than the partial BI scheme and which can be implemented more gradually over time can be proposed.

4. AN ALTERNATIVE ROUTE

The alternative route to be outlined makes use of three elements: the tax allowance (V), the net minimum wage (M) and the withdrawal rate (k) (the poverty trap). By carefully choosing each of these variables it is possible to attain a gross-net traject which closely resembles the gross-net traject of a BI regime. The first two steps are undertaken for the alleviation of the poverty trap and a reduction of minimum wage labour costs respectively, and they can be implemented even if one does not want to implement a BI. This is not true for the third step.

The transition problem is how the move from SAC to SD should be accomplished (see Chart 1). Chart 3 shows the first ingredient, a withdrawal rate less than 100%.

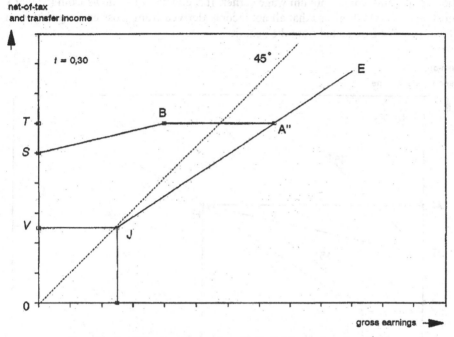

Chart 3 – Conditional social security with limited retained earnings (ST)

Suppose that the government decides that the unemployed are allowed to retain part of their earnings (say, 25%) on top of their benefit up to a maximum of ST. The line which represents all possible combinations of gross and net income is SBA"E. The poverty trap is reduced from below and has become shorter (the distance BA" in Chart 3 is shorter than SA in Chart 1, although the distance SA" measured along the horizontal axis is longer than SA). The government may also choose a withdrawal rate which, given the level of the net minimum income M and the social minimum S, ensures that those receiving benefits and performing (part-time) work can never attain a net income higher than the net minimum income of a full-time worker. This is depicted in Chart 4 by the line SHK.

Given S, M and t the withdrawal rate k can be determined by the following formula:[9]

---

[9] If we take the withdrawal rate k over *net* earnings, formula (5) becomes:
$S + (1-k)[(y_m - V)(1-t) + V] = M$ with $y_m = (M - tV)/(1-t)$, so $k = S/M$.

$$S + (1 - k)y_m = M = k = (y_m - (M - S))/y_m \tag{5}$$

with $y_m=(M-tV)/(1-t)$. Equation (5) states that, given the social minimum is a fixed percentage of the net minimum wage ($S = fM$), the closer the social minimum to the net minimum wage (a higher f), the higher the withdrawal rate k must be in order to keep the net income of those receiving any positive amount of benefits below the net income of a full-time minimum wage earner. If S equals M it follows from (5) that k must equal 1, which means that all net income derived from gross earnings between O and y_m will be deducted from the benefit.

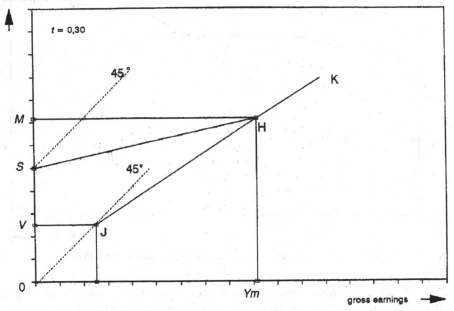

Chart 4 – Conditional social security with maximum retained earnings

Meanwhile, the purpose of this whole exercise (to transform the curve SAC to SD) must not become lost. One has to keep in mind that in a pure BI regime there is no minimum wage, no poverty trap and all adults receive the same basic grant. Moreover, all will face the same marginal tax rate as there is just one single proportional tax rate operational in the pure form of the BI. The position of those who are neither performing paid work nor entitled to a conditional social security benefit (mainly dependent housewives) is not changed at all by the change in the state of affairs depicted in Chart 1 to those depicted in Chart 4: they are still at point

---

The expression between accolades is the net income after taxes from which the recipient may retain a share of $(1-k)$ on top of his benefit S.

O. If the previous scenario operated they would receive a partial BI (they would move to point G in Chart 2), but until now nothing has been done to change their conditions as they are not subject to the poverty trap. So far, the poverty trap has been made less severe. For those receiving a benefit the poverty trap is attenuated by choosing a withdrawal rate less than 100%, but this rate will be higher than the normal tax rate t for workers with labour incomes above the gross minimum wage. If this policy measure is not self-financing, there will be an increase in the gross minimum wage.[10]

Although welfare recipients are now allowed to keep part of their earnings, their chances to find employment are still meagre due to the relatively high gross minimum wages. All potential jobs with a productivity between zero and the minimum hourly wage rate will not be undertaken. For employers it is not profitable to hire unemployed who have a productivity level which falls short of the gross minimum wage. Even if potential employees want to work against an hourly wage rate which lies below the minimum wage rate this is forbidden by the minimum wage legislation. And probably, the density of those with relatively small productive assets or human capital is particularly high among the unemployed.

The above analysis suggests that the next ingredient in the transformation process is to look for ways of lowering the gross minimum wage. The difficulty we face here is that the social minimum S is tied to the net minimum wage M (see Equation (2)). If we keep the tax rate and the tax allowance at the same levels, a reduction in the gross minimum wage rate will not only lower the net minimum wage, but also lower the social minimum, which is undesirable for maintaining the effective guarantee of a minimal share in welfare. There are, however, two other ways of lowering the gross minimum wage, while keeping the levels of S and M unchanged: One could lower the tax rate for earnings up to the gross minimum wage or increase the tax allowance. This can be seen from the general expression of the gross minimum wage given by Equation (2a) above. If we set t equal to zero for earnings up to the gross minimum wage or raise the tax allowance V to M there will no longer be a difference between gross and net minimum wages, which means that the gross minimum wage can be reduced down to M. In fact, the choice between a lower tax rate for the first income bracket or a higher tax allowance is not a real one, as the two are equivalent. For the minimum wage earners it makes no difference to their tax liabilities (or the marginal tax rate) if the tax allowance is made equal to the minimum wage or if the tax rate for earnings up to the minimum wage is set equal to zero.

Of course both measures will cause a decline in the amount of tax revenues. Abstracting from dynamic effects on the economy it is possible to derive the increase in the tax rate that is necessary to keep the total amount of tax revenues (T) constant (i.e. the tax rate that is applied to earnings above the gross minimum wage). By using V_0 for the old tax allowance and V_1 for the new tax allowance, and assuming n workers all with gross earnings greater than V and on average equal to $\overline{y}$, the new tax rate t' can be derived from the following equations:

---

[10] If the reduction in the withdrawal rate to less than 100% is not self-financing, the tax rate must increase which will cause an increase in the gross minimum wage, given the level of the net minimum wage.

$$T = nt(\bar{y} - V_0) = \quad t = T /[n(\bar{y} - V_0)]. \tag{6a}$$

$$T = nt'(\bar{y} - V_1) = \quad t' = T /[n(\bar{y} - V_1)]. \tag{6b}$$

The formulas (6) state that tax revenues are proportional to the total amount of gross earnings above the tax allowance. Dividing (6b) by (6a) gives:

$$t' = t(\bar{y} - V_0)/(\bar{y} - V_1) \tag{7}$$

For a given increase in the tax allowance, the corresponding necessary increase in the tax rate is lower the higher the average income.[11] As both measures are equivalent, Equation (7) can also be interpreted as the new tax rate which will result if the tax rate for earnings up to V_1 is set to zero.

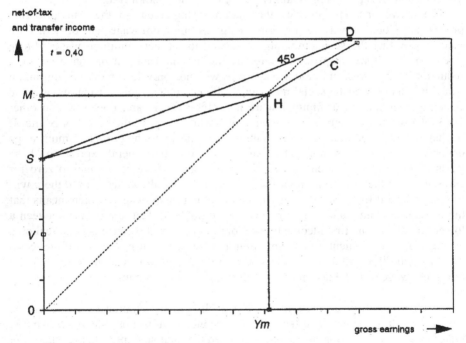

Chart 5 – Tax allowance equal to the minimum wage

---

[11] Consider two countries with equal average tax rates (0.35) and equal tax allowances (500), but different average incomes (2000 or 4000). If both countries decide to raise the tax allowance to 1000, the new tax rate according to (7) in the low-income country will increase to 0.525 and in the high-income country to 0.41.

Suppose that the government chooses to reduce the tax rate on family income at or below the gross minimum wage to zero (thus $V_1 = M$), while this tax reduction in the income bracket $[0, y_m = M]$ is accompanied by a rise in the tax rate for all family income above y_m in order to keep tax revenues at a constant level. Substituting $V_1 = M$ in (7) gives:

$$t' = t(\bar{y} - V_0)(\bar{y} - M)].\tag{8}$$

The withdrawal rate for welfare recipients can again be derived from the condition that their net income may not exceed the net minimum wage:

$$S + (1 - k)M = M \Rightarrow k = S / M.\tag{9}$$

This situation is depicted in Chart 5. The withdrawal rate k is now further reduced to f (given that $S = fM$). The line SHC is now very close to the line SD which corresponds to the pure BI scheme, and it may even be more close to a 'claw-back' BI regime with a surcharge.[12]

Within a structure of conditional benefits which incorporates withdrawal rates and a linking of social security benefits to the net minimum wage, there are no further possibilities to reduce the gross minimum wage without seriously destroying monetary incentives to work.[13] From now on the only possibility to further reduce the gross minimum wage is to introduce a partial BI (G), which partly replaces the tax allowance M. This is illustrated by Chart 6.[14]

If we choose a partial BI equal to the difference between the old net minimum wage and the social minimum ($G = M-S$), the (net and gross) minimum wage (and the tax allowance) can be reduced by the amount G. Someone with gross earnings equal to the new minimum wage (S) has a net income of ($G+S$). The withdrawal rate is now equal to ($S-G$)/S. If G is increased further, the minimum wage moves further

---

[12] A claw-back BI regime is a two-tier regime which has two tax rates. The high rate (t_h) in the first income bracket is such that the maximum amount of taxes paid over this bracket equals the BI (G): $t_h y_d = G$.
All gross earnings above y_d are taxed at the lower tax rate t_l. This means that one pays the high tax rate as long as one receives more in the form of a BI than one pays in the form of taxes. The surcharge is the difference between the high and the low tax rate ($t_h - t_l$). A two-tier BI regime is sometimes put forward in order to reduce the disincentives of a high marginal tax rate for high-income earners.
[13] One possibility left is to lower M accompanied by an increase in f (given $S = fM$) in order to keep the social minimum at the same level. But this possibility will bring the net income of a full-time minimum wage worker very close to a full-time welfare recipient. This may reduce their monetary incentives to (find) work and it will reintroduce the poverty trap.
[14] One can think of an intermediary step between the schemes depicted in Charts 5 and 6. Instead of introducing a partial BI to all adults it is possible to discharge social security recipients of their duty to search for work or to accept work. If one can prove that (family) income is below the social minimum, one can claim a social security benefit. This amounts to a state-supported right to be idle.

towards the origin.[15] But a higher G will induce a higher tax rate in order to keep the budget balanced, which will cause the line HC to rotate clockwise towards SD.[16]

Alternatively, as soon as a partial BI scheme is implemented, we can anticipate the full BI regime effects. Suppose that we know what someone's total net income would be with earnings equal to y_m under a full BI regime[17] (see point H in Chart 6). Given point H we can derive the level of the partial BI which ensures that the first section (SH) of both a full BI regime and a partial BI cum tax allowance equal to the minimum wage coincides. The line HC is steeper than HD because the tax rate in the first regime is probably higher than the latter. Further increases in G mean that the minimum wage earner ends up with a net income somewhere on the line SH. If the partial BI approaches S, the minimum wage approaches zero and the marginal worker ends up with a net income equal to S, a full BI.

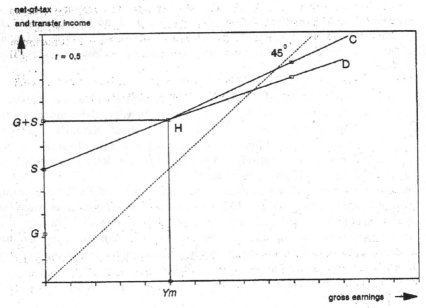

Chart 6 – Tax allowance equal to the minimum wage combined with a partial basic income

The alternative route outlined above consists of three steps. Firstly, allow welfare recipients to keep part of their earnings on the condition that their net income may never exceed the net income of a minimum wage earner. Secondly, raise the tax

---

[15] Given the level of S, a higher partial BI will lower the minimum wage and the tax allowance at the same time.

[16] Dependent on the levels of G and the net (total) income of a minimum wage worker the withdrawal rate can be higher or lower than the tax rate. Suppose we can find a partial BI for which the withdrawal rate equals the tax rate. The line SHC will then be a straight line and further increases in the partial BI can be made which means that the net income of a full-time minimum wage worker will lie on the line SH.

[17] That is, we know what the balancing budget tax rate is in a full BI regime.

allowance up to the level of the net minimum wage which lowers the gross minimum wage to the net minimum wage M. Up to this point the position of those neither performing paid work nor receiving benefits (group 1) is not changed. Thirdly, substitute a partial BI for the tax allowance. Only with this last step will group 1 move from O to G. The unemployed remain at point S, but if the minimum wage is lowered, their chances to find employment will increase which means that more recipients of conditional social security will be on the line SH. Any further increase in the partial BI gives rise to an equivalent decrease in the minimum wage until it can be abolished altogether when the full BI stage is reached.

5. PART-TIME WORKERS

So far I have scarcely paid attention to the position of part-time workers. If the ultimate aim is to move forward towards a full BI regime, it is necessary that during the transition process their position too should move closer to what it would be under a full BI. Using the figures, it is easy to describe the changes in position which occur when different schemes are in operation.

In Chart 1 part-time workers are located on the line OJAHC, with relatively few of them on the section HC. A direct implementation of a partial BI will cause part-time workers to be located on the line GA'H'C in Chart 2. What was previously said concerning persons belonging to group 1 equally applies to part-time workers. They are not lacking the means of subsistence, therefore, from the perspective of a minimum income guarantee there is no reason to give part-time workers, of whom many have a working partner, an unconditional benefit from the state. The introduction of a partial BI will cause the tax rate to rise. If many of them decide to abstain from part-time work under a partial BI regime the tax rate will rise even further. This can be seen more clearly in Chart 1. Whereas part-time workers would be located on the line OJAC under conditional social security, under a full BI regime they are on the line SD. When many of the part-time workers on OJAC are complementing a family income, they have the real choice of abstaining from part-time work altogether once a BI of S is introduced.

In the alternative route the sequence of changes in the tax-and-transfer system is different. The first steps along the alternative route are only meant to strengthen the position of the least advantaged members of society (by allowing them to retain part of their earnings on top of their benefit and by reducing the gross minimum wage, which will increase labour demand at the bottom end of the labour market). Of course, some of the part-time workers would be members of the least advantaged group, if they would not perform part-time work (the same applies for many of the low skilled full-time workers). But this only strengthens the force of the alternative route as it delays policy measures which would probably reduce incentives to work for those who have enough means for an adequate livelihood. Only in the final stages of the transition process towards a BI do these people see any improvement. The least advantaged see the positive effects first.

In Chart 4 the poverty trap is eliminated for those who are worst off. Part-time workers are still located on the line OJHK and rightly so because most part-time workers do not face the poverty trap. In so far as they do, they will be on the line SH. In Chart 5 the gross-net earnings trajectory is made more favourable for part-time workers without a working partner. They do not have any tax liabilities up to the gross earnings level of a full-time minimum wage worker. But part-time workers with a working partner have to share the tax allowance y_m with their partner (since taxation is based on family income). This means that up to the point half-way between O and H they move along the 45-degree line, and thereafter they will move along the line which runs parallel to HC. Due to the decrease of the gross minimum wage the greatest advantage for part-time workers is probably more employment opportunities at the lower end of the labour market for which they have to compete with social security recipients who are allowed to retain part of their earnings. A really significant improvement in their position occurs if a partial BI is introduced (Chart 6). Part-timers without a working partner move along the line GHC, while those with a working partner have to share the tax allowance (they will be located on the first half of the line GH, and after they have used up half the tax allowance on the line parallel to HC). If the partial BI is increased further, their gross-net earnings trajectory will move closer to the line SD, which describes a full BI regime.

6. A DIFFERENTIAL BASIC INCOME

The phase has now been reached where one can aim for a full BI, but we have to deal with the question of whether it should be indeed on an individual basis or not. At present, welfare arrangements still bear the vestiges of its orientation on the traditional family with a single breadwinner, many social security provisions are not unconditional with respect to the wealth and income of the living unit, and many women without an income of their own but with a working partner are not entitled to a social assistance benefit. An individualized social security system, which is unconditional with respect to the circumstances of the living unit, is more in line with the principle of equal treatment, but alas, it is more costly. If one conceives the household or family as units of income and expenditures it is natural, in order to attain effective minimum income guarantees, to make social security benefits conditional on household or family composition.

Now a BI is unconditional in several respects, and being unconditional with respect to family, household or living unit composition is just one of them. There are some important reasons why one should advocate a *differential* BI, dependent on household composition. Firstly, the most important aim of the present welfare state is the effective guarantee to all citizens of the satisfaction of basic needs and a minimal share in welfare. To individualize this right through a BI means that the costs of providing this guarantee will be much higher than the costs of a BI which takes the household composition into account. As a consequence, the level of the individualized BI will be lower for single person households than that received in a differential BI scheme. Secondly, it is questionable whether a differential BI, which is dependent on household composition, violates the principle of equal treatment, given that economies of scale in consumption within a household is a well established empirical fact.[18] A differential BI reflects this empirical fact. Policymakers here face a conflict between the principle of equal treatment and the aim of providing effective social security. Thirdly, the incidence of poverty is especially strong among one-income households. There is a serious danger, because of the decline of the single breadwinner family and the rise of two-earner families, that a great part of future income inequality will be between one and two-earner families. Finally, the rationale behind a differential BI is not to discriminate between one and multi-persons households, or to promote one or the other, but to maximize the guaranteed social minimum. As soon as somebody leaves the household, he or she is entitled to a differential BI for a single person.

To see the difference, let N be the number of adults entitled to a BI, B the total amount available for providing a BI, s the fraction of single person households and t the fraction of two person households.[19] Let b_s be the BI for a single-person household and let b_t the total sum of BI received in a two-person household. So we have:

$$s + t = 1, \quad s, t \geq 0; \tag{10}$$

$$b_t = f b_s, \quad 1 \leq f \leq 2. \tag{11}$$

If f equals 2, it does not matter whether one lives in a one- or a two-person household. If f equals 1, all households, irrespective of their composition, receive the same BI.

First consider the case of an individualized BI ($f = 2$). In that case we have just N persons all receiving the same BI b. Therefore,

$$b = B/N. \tag{12}$$

Next consider a differential BI with f between 1 and 2. Thus,

$$s b_s N + t b_t N/2 = B. \tag{13}$$

Substitute (11) in (13) which gives:

$$s b_s N + t f b_s N/2 = B. \tag{14}$$

Rearranging (14) gives:

$$b_s = B/[N(s + t f/2)]. \tag{15}$$

---

[18] See the *Journal of Income Distribution* 4 (2), 1994/1995 for a theoretical discussion and empirical estimates of equivalence scales. For a BI we ideally should have data on equivalence scales for different families or households for all expenses on commodities which belong to a package which is classified as the social minimum.

[19] For simplicity I assume that there are only one and two-person households.

Dividing (12) by (15) and substituting (10) gives:

$$b/b_s = [s(2-f) + f]/2. \tag{16}$$

Equation (16) expresses that b_s equals b only in the trivial cases where f equals 2 (which corresponds to an individualized BI) or if s equals 1 (which corresponds to the case where all households are one-person households). If f is greater than 1 and less than 2, b_s is greater than b. The smaller the fraction of single person households, the greater the deviation of b_s from b. To make up our minds, if f equals 1.5 and s equals 0.2 (the actual figure in The Netherlands), the BI in a differential regime can be 25% higher for a single-person household compared to an individualized BI. To put it as simple as possible, with a fixed amount of money (B) available for providing a BI to all, giving a two-person household an amount that is less than proportional to their number is equivalent to giving a one-person household an amount that is more than proportional. The effect is larger the smaller the fraction of one-person households.[20] Alternatively, one can easily derive that the reduction in expenses by providing a BI differentially instead of individually equals $[t(1-f/2)]100\%$.

SUMMARY AND CONCLUSIONS

In this chapter an alternative route towards a BI is proposed. The first two steps, to allow welfare recipients to retain part of their side-earnings and to lower the gross to the net minimum wage by raising the tax allowance to this level, can be taken independently whether one aims toward a BI or not. However, the important point to be stressed is that there are good reasons to start with these two steps and not with a partial BI (the third step). The two most important reasons are that the alternative route is more gradual in the sense that it departs entirely from the basic features of social security provided by the conditional welfare state and that it eliminates two shortcomings of that system: the poverty trap and the relatively high gross minimum wages. Of course, any proposal has winners and losers and efficiency gains and losses. The winners are mainly those entitled to social assistance benefits, the losers are mainly those who are neither entitled to these benefits nor perform any paid work, and high income earners. Along the road towards a full BI efficiency gains can be reaped from the elimination of the poverty trap and from the positive dynamic effects of the reduction of the level of the gross minimum wage. Losses are to be expected in so far as these measures cause the tax rate to rise. A real appreciation of these gains and losses and of the effects on the income distribution requires a general equilibrium framework which takes welfare and taxation arrangements into account. Admittedly, the analysis here is comparative and static.

---

[20] At present, income tax liabilities are dependent on family income. If one gives all adults the same individual BI b, one can still get the same result brought about by a differential BI regime by means of taxing two-person households' family income between ($2b - fb$) at a 100% tax rate. The advantage is that one can make use of information already available at the tax authorities.

The BI proposal has, from an efficiency point of view, at least two important strings on its bow: the removal of the poverty trap and the abolition of minimum wages. Part of these efficiency gains will reveal themselves during the first two steps. Thus, despite the serious limitations connected to such a static approach as adopted here, if it turns out that the first two measures proves too costly, it will become very difficult to implement a BI anyway, regardless of the route chosen.

CONCLUSION

In all welfare states the debate on welfare reform continues, especially the form that welfare arrangements should take. Roughly we can discern a shift from a passive to a more active welfare state. Passive welfare merely provides income support as compensation for all kinds of contingencies (illness, disability, unemployment). The active welfare state tries to reduce unemployment by tightening the work-test: from a duty to apply regularly for jobs to a more demanding policy of asking reciprocal efforts from welfare recipients in return for the social benefit. This policy avoids that welfare recipients have to wait and see whether a job comes along and simply stay at home. Instead, mandatory participation in (re)training, public employment or subsidised private employment programmes is put forward as a promising effort to combat social exclusion and poverty, and also motivated by striking the balance between social security rights and its correlative duties. This book advocates to go into the opposite direction, that of an unconditional basic income (BI). A positive justification for the proposal of a substantial BI is given, taking into account the majority support (opposition) for workfare (BI) and the serious objections (parasitism, exploitation) made in the literature. A substantial BI may adapt the postwar institutions of social security to the dynamism of contemporary labour markets, combine adequate social protection with the demands of flexible labour markets, it may lead to a more equitable distribution of income, of paid work, of care work and of free time between men and women, to less involuntary unemployment, better working conditions for low wage workers and to a reduction in administrative costs for providing minimum income social security.

In the first chapter the idea of BI was confronted with three popular notions of justice: self-reliance, reciprocity and the work ethic. I concluded that the idea of a BI is not in apparent conflict with self-reliance. To some extent a BI may even increase the degree of self-reliance among the least productive workers when compared to a work- and means-tested welfare scheme. Although the disbursement of an unconditional BI does not fit nicely into the government-propagated perfectionist work ethic, I argued that promoting this type of work ethic is not a proper task of a liberal government (committed to non-discrimination among various conceptions of the good life). However, the demands of reciprocity and the neutral work ethic pose a very serious problem, namely of parasitism. Taking the reciprocity-based parasitism objection seriously clearly imposes severe restrictions on any strategy which tries to justify a substantial BI.

In the next two chapters the plea for a BI is made against the background that by and large unemployment appears to be a permanent phenomenon in modern capitalist welfare states. In chapter 2, I have shown that attaining compensatory justice on the labour market, especially at the bottom end, is greatly facilitated by the provision of a substantial BI. This argument is particularly relevant in the circumstance of involuntary unemployment, which can be considered as the primary threat to compensatory justice. In this respect, parasitism under the BI scheme (that is, to allow voluntarily unemployed persons to be on welfare) can be seen as a price to be paid to attain compensatory justice. Chapter 3 shows that, under the assumptions made, there is a justification for a BI under less than full employment. If market economies are stuck with the problem of unemployment, and want to maintain social security and equal opportunities for all, then the BI scheme is an attractive alternative. Under unemployment (job scarcity), a substantial BI is warranted and the BI received by those who freely choose not to do paid work can be seen as a compensation for giving up their equal right to jobs.

To show the justice of a BI, however, is not enough. It may be that although a particular type of (e.g. household based) BI, under specific circumstances (unemployment), is considered as a more just system of minimum social security than the present, conditional scheme, such a BI is economically unfeasible. Also, a BI that is economically feasible may not be just. Even if the BI would engender sheer positive economic effects, it has to be shown that it is more in line with the demands of justice than alternative schemes of social security. There is therefore no priority of one dimension above the other. Still, both dimensions are not independent of each other. For example, in chapter 2, I argued that a BI scheme is more conducive to compensatory justice, provided that the economically sustainable BI is not far below the level of the present conditional social benefits. Not much is said about whether or not a substantial BI is economically feasible. To reduce the uncertainty around the feasibility of BI, it would be a good idea to conduct a real life experiment. The outline of a proposal for such an experiment was presented in chapter 4. If such an experiment would show encouraging results with respect to the feasibility of a BI, then one might consider to transform the present conditional system of social security into an unconditional system of minimum income support. The steps to be taken to move gradually from the present system towards a BI system were outlined in chapter 5.

REFERENCES

Arneson, R.J. (1990), Is Work Special? Justice and the Distribution of Employment, *American Political Science Review, 84* (4), 1127-1147.

Arrow, K.E. (1973), Some Ordinalist-Utilitarian Notes on Rawls's Theory of Justice, *The Journal of Philosophy, LXX* (9), 245-263.

Atkinson, A.B. and H. Sutherland (1988), *Tax-Benefit Models*, ST/ICERD Occasional Paper no. 10, London School of Economics.

Atkinson, A.B. (1993), Participation Income, *Citizens Income Bulletin, 116*, 7-11.

Atkinson, A.B. (1995a), *Public Economics in Action. The Basic Income/Flat Tax Proposal*, Oxford: Clarendon Press.

Atkinson, A.B. (1995b), *Incomes and the Welfare State: Essays on Britain and Europe*, Cambridge: Cambridge University Press.

Atkinson, A.B. (1995c), On Targeting Social Security: Theory and Western Experience with Family Benefits, in D. Walle and K. Nead (eds.), *Public Spending and the Poor: Theory and Evidence*, Baltimore-London: John Hopkins University Press, 25-68.

Atkinson, A.B. (1996), The Case for a Participation Income, *Political Quarterly, 67* (1), 67-70.

Baker, J. (1992), An Egalitarian Case for Basic Income, in Ph. Van Parijs (ed.), *Arguing for Basic Income*, London-New York: Verso, 101-127.

Barry, B. (1992), Equality Yes, Basic Income No, in Ph. Van Parijs (ed.), *Arguing for Basic Income*, London-New York: Verso, 128-152.

Barry, B. (1997), The Attractions of Basic Income, in J. Franklin (ed.), *Equality*, London: Institute for Public Policy Research, 157-171.

Beer, P. de (1987), De bezwaren tegen het basisinkomen gewogen, *Socialisme en Democratie, 44*, 50-58.

Beer, P. de (1993), *Het verdiende inkomen*, Wiardi Beckman Stichting, Houten: Bohn Stafleu Van Loghum.

Besley, T. (1990), Means Testing Versus Universal Provision in Poverty Alleviation Programmes, *Economica, 57* (225), 119-129.

Borghans, L. and L.F.M. Groot (1998), Superstardom and Monopolistic Power: Why Media Stars Earn More Than Their Marginal Contribution to Welfare, *Journal of Institutional and Theoretical Economics (JITE), 154* (3), 546-571.

Brown, W.S. and C.S. Thomas (1994), The Alaska Permanent Fund: Good Sense or Political Expediency?, *Challenge*, 38-44.

Burtless, G. (1990), The Economist's Lament: Public Assistance in America, *Journal of Economic Perspectives, 4* (1), 57-78.

Burtless, G. and J.A. Hausman (1978), The Effect of Taxation on Labour Supply: Evaluating the Gary NIT Experiment, *Journal of Political Economy 86* (6), 1103-30.

Byrne, S.E. (1993), *A Rawlsian Argument for Basic Income*, Dublin, National University of Ireland, University College, MA Thesis.

Card, D. and A. Krueger (1995), *Myth and Measurement: The New Economics of the Minimum Wage*, New Jersey: Princeton University Press.

Carens, J.H. (1985), Compensatory Justice and Social Institutions, *Economics and Philosophy, 1* (1), 39-67.

Centraal Planbureau (CPB) (1992), *Nederland in Drievoud*, Den Haag: Sdu.

Clark, A.E. and A.J. Oswald (1994), Unhappiness and Unemployment, *The Economic Journal, 104* (424), 648-59.

Cohen, G.A. (1995), *Self-ownership, Freedom, and Equality*, Cambridge: Cambridge University Press.

Creedy, J. (1996), *Fiscal Policy and Social Welfare: An Analysis of Alternative Tax and Transfer Systems*, Cheltenham: Edward Elgar.

Dick, J.C. (1975), How to Justify a Distribution of Earnings, *Philosophy & Public Affairs, 4*, 248-272.

Dolado, J. *et al.* (1996), The Economic Impact of Minimum Wages in Europe, *Economic Policy: A European Forum, 0* (23), 319-372.

Elster, J. (1986), Comment on Van der Veen and Van Parijs, *Theory and Society, 15*, 709-722.

Elster, J. (1989), *Solomonic Judgements*, Cambridge: Cambridge University Press.

138 REFERENCES

Fitzpatrick, T. (1999), *Freedom and Security: An Introduction to the Basic Income Debate*, London/ New York: MacMillan/St. Martin's Press.

Friedman, M. (1962), *Capitalism and Freedom*, Chicago: University of Chicago Press.

Gelauff, G.M.M. and J.J. Graafland (1994), *Modelling Welfare State Reform*, Amsterdam: North-Holland.

Goodin, R.E. (1988), *Reasons for Welfare: The Political Theory of the Welfare State*, New Jersey: Princeton University Press.

Goodin, R.E. (2001), Work and Welfare: Towards a Post-productivist Welfare Regime, *British Journal Political Science, 31*, 13-39.

Gottschalk, P. (1997), Inequality, Income Growth, and Mobility: The Basic Facts, *The Journal of Economic Perspectives, 11* (2), 21-40.

Greenaway, D. (ed.) (1996), Policy Forum: Economic Aspects of Minimum Wages, *Economic Journal, 106* (436), 637-676.

Groot, L. and R. Van der Veen (2000a), How Attractive is a Basic Income for European Welfare States?, in R. Van der Veen and L. Groot (eds.*), Basic Income on the Agenda: Policy Objectives and Political Chances*, Amsterdam: Amsterdam University Press, 13-38.

Groot, L. and R. Van der Veen (2000b), Clues and Leads in the Policy Debate on Basic Income in the Netherlands, in R. Van der Veen and L. Groot (eds.), *Basic Income on the Agenda: Policy Objectives and Political Chances*, Amsterdam: Amsterdam University Press, 197-223.

Hamminga, B. (1992), Could Jobs be like Cars and Concerts?, unpublished manuscript.

Hamminga, B. (1995), Demoralizing the Labour Market: Could Jobs be like Cars and Concerts?, *The Journal of Political Philosophy, 3*, 23-35.

Heij, J.J. *et al.* (1993), *Basisinkomen in drievoud*, Amsterdam: Het Spinhuis.

Hersog, H.W. and A.M. Schlottmann (1990), Valuing Risk in the Workplace: Market Price, Willingness to Pay, and the Optimal Provision of Safety, *The Review of Economics and Statistics, 72* (3) 463-470.

Howard, M.W. (1998), *Basic Income and Cooperatives*, 7th International Congress on Basic Income (7-9 september 1998, Amsterdam, The Netherlands).

Hwang, H. *et al.* (1992), Compensating Wage Differentials and Unobserved Productivity, *Journal of Political Economy, 100* (4), 835-858.

Keeley, M.C., Ph. K. Robins, R.G. Spiegelman and R.W. West (1978), The Estimation of Labor Supply Models using Experimental Data, *American Economic Review 68* (5), 873-87.

Kennan, J. (1995), The Elusive Effects of Minimum Wages, *Journal of Economic Literature, 33* (4), 1950-1965.

Kershaw, D. and J. Fair (1976), *The New Jersey Income-Maintenance Experiment: Operations, Surveys, and Administration, Vol. I*, New York: Academic Press.

Killingsworth, M.R. (1986), A Simple Structural Model of Heterogeneous Preferences and Compensating Wage Differentials, in R. Blundell and I. Walker (eds.), *Unemployment, Search and Labour Supply*, Cambridge: Cambridge University Press, 303-317.

Klamer, A. and T. Van Dalen (1996), *Telgen van Tinbergen: Het verhaal van de Nederlandse economen*, Amsterdam: Balans.

Lenkowski, L. (1986), *Politics, Economics, and Welfare Reform: The Failure of the Negative Income Tax in Britain and the United States*, Washington D.C.: University Press of America.

Levine, A. (1995), Fairness to Idleness: Is There a Right not to Work?, *Economics and Philosophy, 11* (2), 255-274.

Masters, S. and I. Garfinkle (1977), *Estimating the Labor Supply Effects of Income-maintenance Alternatives*, New York: Academic press.

McKinnon, C. (2003), Basic Income, Slef-Respect and Reciprocity, *Journal of Applied Philosophy 20* (2), 143-158.

Mill, J.S. (1968), Principles of Political Economy with some of their Applications to Social Philosophy, in J.M. Robson (ed.), *Collected Works of John Stuart Mill, Vol. II*, London: Routledge.

Moynihan, D.P. (1973), *The Politics of a Guaranteed Income: The Nixon Administration and the Family Assistance Plan*, New York: Random House.

Munnell, A.H (ed.) (1987), *Lessons from the Income Maintenance Experiments*, Boston: Boston Federal Reserve Bank.

Murray, Ch. (1984), *Losing Ground, American Social Policy 1950-1980*, New York: Basic Books.

Nooteboom, B. (1993), Het basisinkomen als basis voor ondernemerschap, *ESB, 78*, 946.

Peck, J. and N. Theodore (2000), 'Work First': Workfare and the Regulation of Contingent Labour Markets, *Cambridge Journal of Economics, 24*, 119-38.

Plant, R. (1993), Free Lunches Don't Nourish: Reflections on Entitlements and Citizenship, in G. Drover and P. Kerans (eds.), *New Approaches to Welfare Theory*, Aldershot: Edward Elgar, 33-48.

Purdy, D. (1990), *Work, Ethics and Social Policy: A Moral Tale*, Third International Conference on Basic Income (september 1990, Florence, Italy).

Rawls, J. (1971), *A Theory of Justice*, London: Oxford University Press.

Rawls, J. (1982), Social Unity and Primary Goods, in A.K. Sen and B. Williams (eds.), *Utilitarianism and Beyond*, Cambridge: Cambridge University Press.

Rees, A. (1975), Compensating Wage Differentials, in A.S. Skinner and T. Wilson (eds.), *Essays on Adam Smith*, Oxford: Clarendon Press, 336-349.

Robeyns, I. (2000) Hush money or Emancipation Fee: A gender Analysis of Basic Income, in R. Van der Veen and L. Groot (eds.), *Basic Income on the Agenda: Policy Objectives and Political Chances*, Amsterdam: Amsterdam University Press, 121-136.

Roebroek, J.H. and E. Hogenboom (1990), *Basisinkomen alternatieve uitkering of nieuw paradigma?*, s'Gravenhage: Ministerie van Sociale Zaken en Werkgelegenheid.

Rosen, S. (1986), The Theory of Equalizing Differences, in O. Ashenfelter and R. Layard (eds.), *Handbook of Labor Economics, Vol. 1*, Amsterdam: North-Holland, 641-692.

Schmidtz, D. and R.E. Goodin (1998), *Social Welfare and Individual Responsibility*, Cambridge: Cambridge University Press.

Sociaal en Cultureel Planbureau (SCP) (1996), *Sociaal en cultureel rapport 1996*, Rijswijk: SCP.

Solow, R. (1998), Guess Who Pays for Workfare, in A. Gutmann (ed.), *Work and Welfare*, Princeton: Princeton University Press.

Smith, R.S. (1979), Compensating Wage Differentials and Public Policy: A Review, *Industrial and Labor Relations Review, 32* (3), 339-351.

Smith, A. (1982), *The Wealth of Nations*, Middlesex: Penguin.

Snower, D.J. (1994), Converting Unemployment Benefits into Employment Subsidies, *American Economic Review, 84* (2), 65-70.

Tobin, J. (1966), The Case for an Income Guarantee, *Public Interest, 4*, 31-41.

Tobin, J., J. Pechman and P. Miezscowski (1967), Is a Negative Income Tax Practical?, *Yale Tax Journal, 77*, 1-27.

Van der Veen, R.J. and Ph. Van Parijs (1987), A Capitalist Road to Communism, *Theory and Society, 15*, 635-655.

Van der Veen, R.J. (1991), *Between Exploitation and Communism. Explorations in the Marxian Theory of Justice and Freedom*, Groningen: Wolters-Noordhoff.

Van der Veen, R.J. en D. Pels (eds.) (1995), *Het basisinkomen. Sluitstuk van de verzorgingsstaat?*, Amsterdam: Van Gennep.

Van der Veen, R.J. (1998), Real Freedom versus Reciprocity: Competing Views on the Justice of Unconditional Basic Income, *Political Studies, XLVI*, 140-163.

Van der Veen, R. and L. Groot (eds.), *Basic Income on the Agenda: Policy Objectives and Political Chances*, 2000, Amsterdam: Amsterdam University Press (translated into Spanish with additional country surveys on Spain and Latin America as: R. Van der Veen, L. Groot and R. Lo Vuolo (eds.), *La Renta Basica en la Agenda: Objetivos y Posibilidades del Ingreso Ciudadano*, 2002, Buenos Aires: Centro Interdisciplinario para el estudio de Politicas Publicas (CIEPP).

Van Heerikhuizen, B. (1997), Figuraties van zelfredzaamheid, in C. Schuyt (ed.), *Het sociaal tekort: Veertien sociale problemen in Nederland*, Amsterdam: De Balie, 184-193.

Van Parijs, Ph. (1987), *Basic Income, Employment Subsidies and the Right to Work*, unpublished manuscript, Louvain-la-Neuve.

Van Parijs, Ph. (1991), Why Surfers Should be Fed: The Liberal Case for an Unconditional Basic Income, *Philosophy & Public Affairs, 20*, 101-131.

Van Parijs, Ph. (1992), Competing Justifications of Basic Income, in Van Parijs (ed.), *Arguing for Basic Income*, London: Verso, 3-43.

Van Parijs, Ph. (1995), *Real Freedom for All: What (if Anything) Can Justify Capitalism?*, Oxford: Clarendon Press.

Vandenbroucke, F. (2001), *Social Justice and Individual Ethics in an Open Society: Equality, Responsibility, and Incentives*, Berlin: Springer-Verlag.

Watts, H.W. and A. Rees (eds.) (1977a), *The New Jersey Income-Maintenance Experiment, Vol. II, Labor-supply Responses*, New York: Academic Press.

Watts, H.W. and A. Rees (eds.) (1977b), *The New Jersey Income-Maintenance Experiment, Vol. III, Expenditures, Health, and Social Behavior; and the Quality of the Evidence*, New York: Academic Press.

Widerquist, K. (2004), *A Failure to Communicate: What (if Anything) Can We Learn From the Negative Income Tax Experiments?*, unpublished, Lady Margaret Hall, Oxford University.

Weir, M., A.S. Orloff and T. Skocpol (eds.) (1988), *The Politics of Social Policy in the United States*, Princeton: Princeton University Press.

Weitzman, M.L. (1984), *The Share Economy: Conquering Stagflation*, Cambridge: Harvard University Press.

Wetenschappelijke Raad voor het Regeringsbeleid (WRR: Netherlands' Scientific Council of Government Policy) (1985), *Waarborgen voor zekerheid*, Den Haag: Staatsuitgeverij, 49-68.

White, S. (1997), Liberal Equality, Exploitation, and the Case for an Unconditional Basic Income, *Political Studies, XLV*, 312-326.

Whynes, D. (1993), The Poverty Trap, in N. Barr and D. Whynes (eds.), *Current Issues in the Economics of Welfare*, New York: St. Martin's Press, 63-86.

Wolff, J. (1998), Fairness, Respect, and the Egalitarian Ethos, *Philosophy & Public Affairs 27* (2), 97-122.

AUTHOR INDEX